Evergreen

ALSO BY TRENT PRESZLER

Little and Often: A Memoir

Evergreen

The Trees That Shaped America

▲ ▲ ▲

Trent Preszler

ALGONQUIN BOOKS
LITTLE, BROWN AND COMPANY

Algonquin Books of Chapel Hill / Little, Brown and Company
Hachette Book Group
1290 Avenue of the Americas, New York, NY 10104
algonquinbooks.com

First Edition: December 2025

Algonquin Books of Chapel Hill is an imprint of Little, Brown and Company, a division of Hachette Book Group, Inc. The Algonquin Books name and logo are trademarks of Hachette Book Group, Inc.

The publisher is not responsible for websites (or their content) that are not owned by the publisher.

The Hachette Speakers Bureau provides a wide range of authors for speaking events. To find out more, go to hachettespeakersbureau.com or email hachettespeakers@hbgusa.com.

Little, Brown and Company books may be purchased in bulk for business, educational, or promotional use. For information, please contact your local bookseller or the Hachette Book Group Special Markets Department at special.markets@hbgusa.com.

ISBN 9781643756707
LCCN 2025943268

1 2025

MRQ-T

Printed in Canada

For my husband, Breno.

. . . and the trees will last forever.

—CHIEF OSHKOSH

CONTENTS

Evergreen

EVERGREEN *(noun)*: A plant whose foliage remains green and functional through more than one growing season. Especially: any of roughly six hundred species of evergreen conifers—ancient relics of a world before dinosaurs—commonly known as arborvitae, cedar, cypress, fir, hemlock, juniper, pine, sequoia, spruce, and yew. Identifiable by simple anatomy: a straight central trunk, branches evenly spaced in a pyramid shape, needlelike leaves, and scaly brown seed cones. The plant kingdom's equivalent to the city pigeon: resilient, adaptable, unfazed by harsh surroundings, making its home in rock crevices, clinging to sheer cliff faces, thriving in environments inhospitable to most other living organisms. Both rowdy and composed at once: the tree you would want on your side in a bar fight.

EVERGREEN *(adjective)*: Having continual and universal relevance, not confined to a particular event or date. Retaining freshness, interest, or popularity. Enduring. Timeless.

PROLOGUE

The Secret Life of Christmas Trees

THE WIND CUT sideways across the prairie the day my father took me to chop down a tree. I must have been seven or eight. We drove out past the barns until the road turned from gravel to ice and the land opened into silence. It was late December, and snow drifted in thin, dry lines across South Dakota's ocean of grass. My father pointed toward a sparse outcropping of evergreens on a bluff, their black silhouettes stark against the whiteness. They grew in this frigid moonscape, I supposed, because nothing else could.

We climbed up the bluff, past dead tree snags guarded by raging crows. It was frightening yet exciting, my first thrill of entering terra incognita. Having stuffed our burlap sacks with enough greenery and pine cones to garland the dining table, we set about choosing the perfect specimen. I circled the trees, inspecting each one from every angle. No two were alike. The young pines stood motionless, waiting to learn which of them would die.

The expedition felt like more than just finding a Christmas tree. Perhaps it was an initiation into the hidden complexities of nature. To identify a ponderosa pine, by counting its long needles grouped in bundles of three, gave me confidence. I thought I really knew something important about the world.

I chose a prickly brute twice as tall as I was, and it absorbed a shocking number of blows from my father's hatchet before keeling over with a creaking snap. We lugged it like a kill back to the truck through knee-deep snow, its branches rustling in surrender. Every few yards, I stopped, hands on my knees, panting. Not one to abandon any struggle, my father carried on with the determination of a big-game hunter. The tree's resinous scent prodded me to keep going.

By the time we had set up the ice-cold, fresh-smelling tree in a corner of the living room, it was almost time for supper. After pie, we all gathered around it. Removed from its wild habitat, the tree appeared alien, its branches pressed awkwardly against corner walls, its grandeur diminished by the linoleum floor and the popcorn ceiling. The trees always looked smaller inside.

From out of a magic box, my mother produced a collection of homemade ornaments: a cross-stitched Jesus, salt-dough candy canes, and a crocheted angel, soaked in starch, stiff enough to stand on its own. Every year, someone had to climb onto a rickety kitchen chair and risk grave personal injury to crown the tree with that blasted angel. Despite annual promises to finally get it right, the angel's halo always ended up bent painfully sideways against the ceiling. Grandpa let me sip red port wine while the whole broken-necked-angel fracas unfolded.

Our tree became a fairy tale. Bird nests woven of lichen and pheasant feathers shared space with ominous pine cones, fat and sticky with sap. My mother wrapped the base in a cotton skirt for that snowy Bavarian village motif. But the tree's real wonder came from an unlikely vantage: the view from below. I crawled underneath to stare up through its branches.

When I let my eyes go out of focus, the dark pine blurred with the red, blue, and orange of Christmas lights winding through the tree. I liked the house to be quiet for this part. Everything went still, save for the soft pinging of the bimetallic strip inside each incandescent bulb snapping open and closed, blinking erratically, like naked little flames springing to life, burning hot, trembling, then vanishing.

The tree wobbled slightly, and I worried for a moment. But my father had bolted it to the floor—or maybe he had secured it to the wall with baling twine. I had seen him do the job, but it still felt like a mystery. Nevertheless, I figured it probably would not fall on top of me. My sisters dashed in to toss on more tinsel. I heard them but paid no attention. I just lay there, hands clasped behind my head, gazing up.

My earliest memories of wonder were born in the quiet, glowing space beneath those branches. The Polish author Bruno Schulz wrote in the 1930s about the dreamlike strangeness of childhood memories, and how certain images we encounter in youth carry lifelong meaning. "They function like those threads in the solution around which the significance of the world crystallizes for us," Schulz wrote. That Christmas tree became one of those threads: an anchor to childhood, a reminder of something elemental and mysterious.

For years, I assumed Christmas would always feel exactly like that. I could not have imagined how quickly the world, and my own traditions, would change.

Thirty-five years later, I sat at my computer, cursor hovering over the BUY NOW button on a thousand-dollar, seven-foot artificial tree with prestrung LED lights, a revolving base, and a spritz of manufactured pine essence. It was the winter of 2020, a bleak pandemic December. I had not seen my family, or anyone else, for longer than I cared to admit. The house was too quiet. The holidays felt like a rerun I could not escape, and I began to wonder if it would be easier to fake the whole thing. Just order a tree and get it over with. I was already buying groceries and toilet paper online anyway—what was one more thing in the virtual cart?

But something about it stopped me cold.

Suddenly, a ritual deeply ingrained since childhood felt clouded by indecision and anxiety. This once-straightforward tradition—pick a tree, bring it home, hang ornaments—now required a remarkably fraught calculation. Real or fake? Disposable or reusable? Fresh-cut or flame-retardant? Something to compost, or something to store in the attic and drag out again next year? A real balsam fir destined to dry to a crisp in weeks, or a plastic facsimile engineered to outlast us all? The marketplace was endless, as was the debate over cost, sustainability, and convenience. How had our society reached a point where we needed investigative journalists to tell us where our trees come from, and economists to calculate the true cost of holiday cheer?

The more I tried to push these questions away, the more they haunted me, like mischievous specters of lost traditions that multiplied with every opened browser tab. Eventually, I closed my computer and found myself watching *A Charlie Brown Christmas* at two o'clock in the morning, looking for something I could not name.

Halfway through, after Charlie Brown confessed his angst about finding the true meaning of Christmas, Lucy offered her famous diagnosis: "Let's face it. We all know that Christmas is a big commercial racket." She sent Charlie Brown to buy a dazzling aluminum tree that would "fit the modern spirit" of the holidays. The tree lot was filled with pink and blue metallic trees—gaudy fakes that seemed to represent all that was wrong with America. But sitting forlorn among the showy specimens, a spindly little pine clung to its last needles on three feeble branches.

"This little green one here seems to need a home," Charlie Brown told a skeptical Linus. "We'll decorate it and it'll be just right... Besides, I think it needs me."

They brought it back, draped it with Snoopy's doghouse décor, and coaxed it into something hopeful. Sparkling with pride—and, apparently, the true spirit of Christmas—the triumphant little tree left me with a satisfied lump in my throat. Here was a tree that was practically dead, barely standing, yet somehow still alive. A miracle. A gift.

"I never thought it was such a bad little tree. It's not bad at all, really," Linus declared, tucking his cotton blanket around the base and propping it up. "Maybe it just needs a little love."

That line stuck. For decades after Charlie Brown's tiny pine first captured the nation's heart in 1965, it served as a kind of emotional shorthand for Americans clinging to their real-tree tradition. At its peak in the 1980s, twenty-two thousand mom-and-pop Christmas tree farms thrived in every corner of the nation, supplying fresh-cut evergreens to approximately 75 percent of US households every year.

Families made rituals out of trips to local tree farms, bundling up against the chill to wander rows of fir and spruce, each tree appraised for height, scent, and needle density. Trees rode home strapped to car roofs, branches bouncing merrily in the wind.

But beneath this affection, a ruthless competitor was methodically building its own fan base.

Artificial Christmas trees, once novelty items fashioned from dyed goose feathers and toilet brushes, evolved into glossy, prelit Bluetooth-enabled replicas so lifelike they seemed real—minus the nuisance of fallen needles littering the floor. Between 2014 and 2024, the US market for imported Chinese-made plastic Christmas trees doubled, to $2.1 billion, with Christmas decorations and lights contributing another $4 billion. As millions of baby boomers, no longer tethered to their childhood traditions, traded authenticity for artificiality, the market shifted dramatically. By 2023, approximately 75 percent of US households displaying trees had switched to artificial models. Real evergreens were fading from the holiday landscape, replaced by manufactured nostalgia and plastic convenience.

Meanwhile, the real-tree industry faced an existential crisis. From a peak of thirty-three million fresh-cut evergreens sold in 1985, annual sales plummeted to an all-time low of just twelve million trees in 2023. Over those decades, farmers retired without successors willing or able to sustain their seasonal businesses, resulting in the closure of 87 percent of tree farms. According to the US Department of Agriculture, the national stock of Christmas trees dwindled to fewer than three thousand farms, covering an area barely larger than the city of Los Angeles. Living evergreens, with their sticky branches and brittle needles, suddenly seemed too fragile and messy for modern life. Droughts scorched delicate saplings before they could mature. One-click shopping replaced roadside tree stands, and when the pandemic upended holiday rituals and shifted consumer spending habits, it accelerated these long-simmering trends.

The evergreen tree, once a living symbol of hope and joy, had become just another endangered product category—but that was only the beginning. The trees were not simply disappearing from Christmas; they were being stripped of their distinctive shape, scent, and symbolism, repurposed into whatever the marketplace demanded next. From holiday décor and furniture to fuel pellets and firewood, the evergreen was reinvented for maximum yield and minimum friction, tailored to a consumer economy increasingly obsessed with instant gratification.

Yet the significance of evergreens extends far beyond their market value. They occupy a vital place within an economic web connecting every aspect of human existence, influencing the daily lives of all eight billion people on the planet. Evergreens provide shelter and fuel, nourishment and trade, inspiration and myth. They have sparked wars, built industries, and anchored economies, through centuries of human exploration, invention, and folly. They have stood longer than kings, outlived empires, and borne witness to humanity's grandest ambitions and deepest failures. Worshipped and burned, protected and plundered, evergreens reflect who we are and reveal the stories we tell about ourselves. In their branches, we hang our aspirations; in their destruction, we reveal our compromises. They remain both victims and accomplices in the human project.

So, what exactly *is* a Christmas tree?

This book presents a fairly involved answer to that seemingly simple question. Along the way, it also tries to figure out how such a simple question ever became so complicated.

My wager in writing *Evergreen* was that to truly grasp humanity's relationship with Christmas trees, I needed to start at the very beginning and follow the trees themselves on their passage from the earth to our living rooms. Retracing their path took me to tree farms in New York and Oregon, lumber mills in Wisconsin and Maine, cargo ships docked in Washington, cattle ranches on the Nebraska prairie, and majestic redwood groves in California. I donned snake-proof boots to traverse pine plantations in Louisiana and North Carolina, and peeked into suburban Home Depot stockrooms on Halloween, where Christmas trees went on sale before trick-or-treaters had even left their houses.

Uncovering the secret life of Christmas trees carried me across not just physical landscapes but also layers of time, from ancient myths and colonial settlements to today's urgent debates about sustainability and consumer culture. By the end, after chasing threads of connection through unlikely places and old books, the significance of these humble trees became clear:

To understand America, start with the evergreen.

I

A Living Altar

▲ ▲ ▲

IMAGINE FOR A moment if visitors from another planet observed humans buying, selling, and decorating billions of trees every year. They might reasonably mistake us for members of a global evergreen cult. If these extraterrestrials watched as trees vanished abruptly from forests around the world, only to reappear overnight, shimmering in living rooms and shop windows, they would marvel at humanity's arboreal obsession and the sheer devotion required to sustain it. To their alien eyes, every house framed in fir, every ship's mast carved from pine, every page bound into books, and every log burned to ash might appear as offerings to ancient gods. And if, through countless glittering bulbs illuminating these trees, they glimpsed New York City's Rockefeller Center at Christmastime, they would surely believe they had discovered the holiest shrine of an evergreen-worshipping civilization.

Every year just after Thanksgiving, millions of Americans turn their gaze toward a single tree: an evergreen carried like a fallen giant on a flatbed truck through the streets of Manhattan. Crowds gather to watch workers climb its trunk and expertly thread fifty thousand lights through limbs that once knew only wind and snow. Then, on a cold Wednesday evening, in a ritual as predictable as a moonrise, someone flips a switch. The tree blazes to life, casting a sudden, surreal daylight across Fifth Avenue. For a few moments, it eclipses phones, emails, deadlines, everything, and commands the attention of an entire nation.

But this glitzy tradition had humble roots. In 1931, America was sinking deeper into the Great Depression, with unemployment nearing 25 percent. Construction workers laboring over Rockefeller Center's emerging art deco skyscrapers battled fierce winds and bitter cold, their futures as uncertain as their paychecks. Desperate for a hint of joy, they pooled what little money they had to buy an unassuming twenty-foot balsam fir, decorating it humbly with handmade garlands, strings of cranberries, and empty tin cans salvaged from lunch boxes. Set against scaffolding and steel girders, the tree stood not as a spectacle, but a symbol of resilience. Far beyond mere decoration, it was a living altar to hope.

Two years later, that private gesture blossomed into a public ritual when Rockefeller Center erected its first official Christmas tree, this one adorned with hundreds of electric lights. Over subsequent decades, the tree grew steadily taller, brighter, and more elaborate, until it became as essential to the American Christmas as Santa Claus himself.

Today, few people feel the holiday season's pressure more acutely than Erik Pauze, Rockefeller Center's head gardener, who shoulders the responsibility of selecting America's most famous evergreen to tower above the skating rink at Rockefeller Plaza. A soft-spoken horticulturist with meticulous instincts and an ever-present safety helmet, Pauze notices details others overlook: the subtle droop of a branch, the shade of green signaling hidden drought stress. He has spent more than thirty years scouring backyards, small-town parks, and forgotten farmsteads across the Northeast in search of perfection. He drives thousands of miles annually, often following tips scribbled onto napkins or whispered by locals in roadside diners. The ideal tree, he says, must have "personality"—an intangible dignity that commands attention amid Manhattan's steel and glass skyline.

Homeowners sometimes submit candidates, hoping their tree will achieve brief immortality, but Pauze often discovers promising evergreens himself while roaming backcountry roads, keeping a mental tally of future contenders.

"I found the tree in Vestal, New York, when I was on my way to look at another tree, not too far away," he told the *New York Post* in 2023. Springtime gossip had already tipped him off. Locals were whispering that the tree was Rockefeller-worthy. When he knocks on a door, "Sometimes they believe me right away," Pauze said. "And sometimes it's like 'Nah, no way. You're not the guy from Rockefeller Center.'"

Finding the tree is just the beginning. Pauze's crew climbs into its canopy to inspect the structural integrity of each limb, ensuring that the tree can safely support the nine-hundred-pound Swarovski crystal star destined for its spire. For months leading to the tree's late-November debut, the crew monitors its water intake and plots every detail of its harrowing journey into New York City.

On cutting day, an early winter hush blankets the site as the crew gently wraps branches in twine and burlap, careful not to snap a single limb. Cranes hoist the massive evergreen onto an industrial trailer specially engineered to distribute its weight evenly, avoiding strain on bridges and tunnels. The convoy, often led by police escorts and trailed by camera crews, moves through rural back roads and small-town main streets, where crowds snap photos and wave handmade signs reading NEW YORK BOUND! As the tree crawls slowly through tunnels, drivers hold their breath, navigating mere inches of clearance. By the time the seventy-to-one-hundred-foot-tall Norway spruce finally reaches Rockefeller Center, its journey has become a televised event, broadcast live from helicopters, watched by millions eager to witness living history in motion.

The tree-lighting ceremony unfolds with the solemnity and grandeur of a high holiday, as television cameras and commerce assume the roles of priest and congregation. In a typical year, around seven hundred thousand people attend in person, while another seven million watch on-screen. Yet amid the jubilant spectacle of its unveiling, the tree's origins, selection process, and transportation recede into the background, overshadowed by celebrity interviews, raucous singing, and twirling skaters. The evergreen itself, an object of reverence and devotion, becomes almost incidental—a backdrop whose deeper significance is lost somewhere between entertainment and tradition.

By mid-January, the show is over. What began as an icon ends as wood. Most years, the once-grand Rockefeller spruce is carefully dismembered, its trunk milled into planks. Volunteers transport this precious lumber to Habitat for Humanity building sites, where the former celebrity begins its second career as anonymous floorboards, roof trusses, or picnic tables.

Even in death, the Rockefeller Center tree exposes a paradox at the heart of Christmas: the uneasy marriage of capitalism and faith, commerce and mythology. The holiday season is neatly packaged as a fairy tale for children, its pagan origins softened into gentle myths, its rituals distilled into magical fantasies. For many adults, however, the season is more about indulgence than mystery: Parties, feasts, alcohol, and frantic Black Friday stampedes blur the lines between meaningful tradition and empty routine.

The Christmas tree embodies all these tensions. To some, it symbolizes peace and goodwill. To others, it is merely a corporate publicity stunt sprinkled with glitter. A decorated evergreen in the heart of a bustling city may genuinely represent holiday cheer—but so, too, might eggnog, Rudolph the Red-Nosed Reindeer, or mounting credit card debt.

Still, the fleeting presence of Christmas trees every December only heightens their power as seasonal markers and cultural emblems. They are a sweet confection placed on the tongue and savored slowly until, with the turning of the New Year, their flavor dissolves and lingers only as a faint memory for eleven months. Perhaps this ephemeral cycle is precisely why evergreens resonate so profoundly. They mirror our own impermanence: always in motion, always fading. Magical yet melancholy, the trees offer both a celebration and a gentle farewell, rolled into one.

All the more reason, then, to enjoy them while we can.

2

Trees That Time Forgot

▲ ▲ ▲

BENEATH THE GLITTERING spectacle of modern Christmas lies a tree far more ancient than the holiday itself—older than humanity. Long before anyone strung lights or hung ornaments, forests of towering evergreens stood shrouded in mystery, their branches holding secrets. Early humans lingered warily at their edges, feeling both awe and apprehension beneath dense, protective canopies. Yet even these ancient forests had predecessors. Millions of years before humans first stepped cautiously into their shadows, the world belonged to a tree called Archaeopteris—forgotten by time, yet responsible for creating the world we now inhabit.

In the Devonian Period, roughly 385 million years ago, this primitive tree dominated the prehistoric world, growing up to sixty feet tall, with a stiff central trunk and a labyrinthine root system powerful enough to break through rock. Well-preserved fossilized remains discovered in 1871 near Gilboa, New York, revealed Archaeopteris as the pioneering architect of modern tree anatomy, with woody branches, a vascular system for transporting water, and the ability to grow year after year without dying back to the ground. Botanists classify Archaeopteris with the *progymnosperm* plants, the precursors to modern conifers, the proto-evergreens—survivors of deep time, which would one day be draped in lights and worshipped as symbols of fertility. Even the fernlike branches of Archaeopteris resembled the spindly arms of a top-heavy Christmas tree.

Its hulking form endowed Archaeopteris with the power to transform Earth's atmosphere by sequestering carbon and generating oxygen. The forests it created blanketed most of the planet's landmass and marked the first great turning point in Earth's biological history, when air became breathable and land became habitable. Its roots harbored an expanding array of ground-dwelling microbes that accelerated decomposition of dead plant matter and created the nutrient-rich soil necessary for new plant lineages to evolve and flourish.

Then, almost as suddenly as it emerged in the Devonian Period, Archaeopteris went extinct in the Carboniferous Period, likely due to volcanic eruptions. But its influence endured. Upon dying, Archaeopteris collapsed into oxygen-deprived swamps, accumulating layer upon layer into vast deposits of peat, the dense, carbon-rich material formed from partially decomposed vegetation. Immense geological compression and heat over millions of years transformed this peat into coal. The nutrients Archaeopteris released into prehistoric oceans fed phytoplankton and zooplankton blooms, which eventually compressed into crude oil. Today, as coal is dug up and burned for electricity, and crude oil is pumped out to be refined into gasoline and plastics, humans are effectively reversing the carbon storage process that Archaeopteris initiated.

A profound irony lingers today. Coal and oil—the two most widely traded commodities and the fuels powering modern civilization—are essentially fossilized prehistoric Christmas trees, the original ghosts of Christmases past. Our cities glow and our cars hum thanks to sunlight captured millions of years ago by forests now buried deep beneath our feet.

From the evolutionary ashes of Archaeopteris, its lineage did not just replicate its pyramidal evergreen form; it defined survival. During the Permian and Triassic Periods, as the planet swung wildly between ice ages and infernos, new trees took root that would shape the modern world: evergreen conifers. By the time dinosaurs roamed the earth, conifers dominated global forests, thriving in conditions that crushed more delicate ferns and cycads.

Today's Northern Hemisphere woodlands still bear the survivalist fingerprints of these ancient ancestors. Instead of broad leaves that wither under cold and drought, conifers evolved needlelike foliage coated in wax, sharply reducing moisture loss and rendering them nearly impervious to winter's grip. When deciduous forests turn skeletal and fields freeze, conifers flourish.

All conifers are evergreens, though not all evergreens are conifers. "Evergreen" refers to any tree retaining foliage year-round, while "conifer" denotes specifically those nonflowering trees that reproduce through seed-bearing cones within the botanical order *Pinales*. This lineage encompasses about six hundred species, including many familiar trees: arborvitae, cedar, cypress, fir, hemlock, juniper, pine, sequoia, spruce, and yew. Their cones, arranged in Fibonacci spirals, feature tough, woody scales and prickly textures. These traits have prevented all but the most determined herbivores—like the Parasaurolophus dinosaur—from swallowing them whole. By safeguarding seeds across geological epochs defined by catastrophic upheaval, cones helped conifers become dominant long before human civilization arose.

Ancient cultures revered evergreen cones for their unique geometry, recognizing them as symbols of enlightenment, rebirth, and immortality. Egyptian priests carried staffs topped with carved pine cones, and Romans adorned fountains and temples with similar motifs, believing the cones' spiraled patterns concealed mathematical secrets waiting to be deciphered. Some theologians have even speculated that the biblical "fruit" in Genesis—the mysterious food eaten by Eve at the serpent's urging, prompting mankind's expulsion from Eden—was actually a pine cone. Though botanists do not classify cones as fruits, the pine cone nevertheless became the evolutionary emblem of evergreens, a natural blueprint for resilience across geological time. A striking relic of Rome's evergreen fascination survives to this day: the *Pigna*, a thirteen-foot-tall bronze pine cone, originally part of a fountain near the Temple of Isis. Legend holds that it once crowned the Pantheon's oculus. Now resting quietly in a Vatican courtyard, the *Pigna* memorializes an era when evergreen forests seemed limitless.

The conifer's ancient legacy of resilience resonated deeply with prehistoric humans, who instinctively revered symbols of vitality. From pine cones carved into Assyrian temples to fir trees celebrated in Norse mythology, civilizations honored conifers as symbols of fertility, longevity, and divine favor. Eventually, people began bringing evergreens indoors, adorning homes and temples with them as talismans against winter's desolation. The trees' living, green boughs served not merely as decoration, but as a reassurance that life could persist through the bleakest months. What began as a simple survival ritual evolved into elaborate ceremonies, laying the foundations for future wintertime celebrations, both spiritual and secular.

In the Roman Empire, the solstice festival Dies Natalis Solis Invicti ("Birthday of the Unconquered Sun"), founded by Emperor Aurelian in AD 274, encouraged unlimited feasting, public intoxication, and ritualized excess, meant to summon back the light on the shortest day of the year. Another midwinter bacchanal, Saturnalia, dedicated to Saturn, the Roman god of agriculture and renewal, turned society upside down for a week: Enslaved people feasted at tables usually reserved for their enslavers; debts were forgiven; and criminals were granted reprieve. Saturnalia's festive chaos blurred the lines between sacred ritual and earthly indulgence. Although the historical record remains ambiguous, some accounts suggest that sexual rites among all-male gatherings of farmers, held within temples constructed from pine logs, were intended to ensure the fertility of their fields. For a brief interval, the world seemed to suspend its normal rules, echoing an ancient yearning for equality and a release from life's harsh realities.

Beyond state-sanctioned Roman festivals, other midwinter

traditions carried explicitly religious overtones. In the fourth century, devotees of Asherah—a Semitic goddess linked to fertility and renewal, whose name derives from the Hebrew word for "merry"—erected evergreen trees as household altars. Worshippers placed gifts beneath idol trees and enacted rituals of rebirth while decorating the trees with symbols of procreation: red orbs, representing testicles; silver strands, for semen; and yonic evergreen wreaths, indicating the female form. These ancient customs signified reverence for life's continuity through winter's darkness—beliefs that would later be dismissed and reinterpreted by dominant religious authorities.

The Old Testament condemned such rituals outright, reflecting ancient Hebrew anxieties about the allure of neighboring pagan traditions. Jeremiah 10:3-5 says: "For the customs of the people are futile: for one cuts a tree from the forest, the work of the hands of the workman, with the ax. They decorate it with silver and gold; they fasten it with nails and hammers so that it will not topple."

Here began a tension that resonated through the ages. Evergreens embodied contradiction, at once sacred and profane.

As the Roman Empire expanded and its pagan winter rites spread across the Mediterranean, the winter solstice became a shared axis of spiritual devotion. Worshippers of the Greek god Dionysus, the Persian god Mithra, the Greco-Syrian god Adonis, and the Hindu god Surya all staged celebrations fixed to December 25, aligned with the Roman calendar's longest night. The official designation of December 25 as the birth date of Jesus Christ, often mistaken as historical fact, emerged much later. In AD 340, Pope Julius I strategically fixed December 25 as the date of the Nativity, absorbing existing pagan solstice traditions into Christian doctrine. About 250 years later, Pope Gregory instructed the theologian Augustine of Hippo, who was sent to convert pagan England, not to abolish local winter festivals, but to repurpose them. Evergreen trees, wreaths, and boughs—all ancient fertility symbols—remained, but their meanings were subtly recast.

This shift had profound implications beyond theology. In his seminal 1967 paper, "The Historical Roots of Our Ecologic Crisis," Lynn White Jr., a history professor at the University of California, Los Angeles, argued that early Christianity's rejection of pagan animism fundamentally altered humanity's relationship with nature. "Christianity made it possible to exploit nature in a mood of indifference to the feelings of natural objects," White wrote. Once stripped of sacred significance, evergreens became commodities, ready to be harvested without reverence.

Despite their diminished spiritual status, evergreens continued to flourish culturally, adapting to new roles within freshly minted Christian celebrations now tied to December 25. As evergreens gradually transitioned from sacred relic to festive decoration, numerous legends tried to explain and justify the trees' presence. One widely circulated myth attributes the tradition to Martin Luther, the Protestant reformer, who is said to have found symbolic beauty in evergreens, seeing their perpetual color as a testament to the enduring grace of God. Another popular tale dating back to the fifteenth century involves Saint Boniface, a missionary who purportedly interrupted a pagan sacrifice beneath an oak tree by cutting it down; an evergreen promptly grew in its place. A variation of this legend describes Boniface cutting down the evergreen after it had matured and then suspending it upside down to represent the Holy Trinity, a common practice in early Christian communities.

While legends undoubtedly helped popularize the Christmas tree tradition, its true roots lie in medieval Germany, which blended myth and reality to nurture the season's most enduring symbol. The earliest documented instance of a decorated evergreen appeared in 1419 in Freiburg, Germany, where villagers adorned a fir with apples, flour-paste wafers, and gingerbread for the Christmas Eve feast of Adam and Eve. In 1510, merchants in Riga, on the Baltic Sea, set up solstice trees decorated with straw, which they burned at Lent. Early references to these Germanic trees often surfaced in regulations written to

limit their harvest, such as the 1530s ordinance in the Alsatian town of
Ammerschwihr stating that no person "shall have for Christmas more
than one bush of more than eight shoe lengths," and the 1611 ban on
cutting Weihnachtsbaum ("Christmas trees") in Turckheim. Such laws
were intended to prevent merchants from ransacking local forests, and
highlight some of the earliest public concern for preserving evergreens
in the face of growing cultural demand.

Given America's puritanical roots, it is hardly surprising that the
German Weihnachtsbaum took so long to catch on across the Atlantic.
William Bradford, the Pilgrims' second governor, considered Christ-
mas celebrations blasphemous, condemning them as "pagan mockery."
In 1659, the General Court of Massachusetts criminalized Christmas,
decreeing that "whosoever shall be found observing any such day as
Christmas or the like . . . shall pay for every such offence five shillings,
as a fine to the country."

One wonders what Governor Bradford might think today, seeing
contemporary evangelical Christians passionately defend the same traditions he so vehemently rejected. Perhaps he might have appreciated
the irony or simply marveled at how religious symbols shift across
generations, constantly reshaped by evolving beliefs and anxieties.

Colonial New England's stern solemnity persisted until the influx
of German immigrants in the 1800s gradually loosened the Pilgrims'
grip on social norms. When Christmas trees finally did take root in
American culture, they carried more than ornaments and lights on
their branches, but the weight of a young nation still figuring out its
identity.

The ink had barely dried on the US Constitution when the debate
over slavery cleaved the nation into North and South, pro-freedom
and pro-bondage. Amid this turmoil, Christmas trees became contra-
dictory symbols of unity and division, deeply entangling evergreens
with the nation's moral crisis. An abolitionist group called the Massachusetts Anti-Slavery Society saw Christmas as a chance to highlight the hypocrisy of a republic proclaiming freedom while holding

millions in chains. In Boston, Christmas bazaars sprang up selling abolitionist-themed gifts, including handmade children's dolls representing enslaved people holding broken chains, and embroidery samplers with abolitionist slogans. These events scandalized polite society at first. Newspapers dismissed abolitionist gatherings as "promiscuous assemblies," deriding male participants as "Aunt Nancy men." Despite early hostility, the bazaars flourished. Abolitionist Christmas cards became popular. Carolers sang about freedom. And, at the urging of German immigrant Charles Follen, organizers erected evergreen trees decorated with ornaments depicting anti-slavery symbols.

A children's rights advocate, gymnastics coach, and Harvard University's first professor of German language, Follen wanted to re-create the Christmas tradition that had enchanted his childhood in Germany. In 1832, he cut a fir tree near his Boston home, decorated it lavishly, and invited guests to witness this strange new custom. Harriet Martineau, a British author and prominent Victorian social theorist, breathlessly described the scene:

> It really looked beautiful; the room seemed in a blaze, and the ornaments were so well hung on that no accident happened, except that one doll's petticoat caught fire. There was a sponge tied to the end of a stick to put out any supernumerary blaze, and no harm ensued. I mounted the steps behind the tree to see the effect of opening the doors. It was delightful. The children poured in, but in a moment every voice was hushed. Their faces were upturned to the blaze, all eyes wide open, all lips parted, all steps arrested.

Perhaps in erecting that tree, Follen found solace in memories of the homeland he had left behind. Or perhaps it was a subtle act of defiance, an evergreen rebuke to a society unwilling to confront its deepest hypocrisies. Though Follen remains largely a footnote in abolitionist history, he deserves recognition as the progenitor of America's evergreen tradition and the man who forever linked it with the nation's struggle over slavery.

Not every Christmas celebration carried such weighty symbolism. Around the same time that Martineau lavished praise on Follen's tree in Boston, newspapers in Philadelphia gleefully reported on the Society of Bachelors' intention to erect a Krischtkintkle Bauhn ("Kriss Kringle tree" in Pennsylvania Dutch) for the amusement of visitors, promising unmatched delight from decorations described as "superb, superfine, superfrostical, shnockagastical, double refined, mill'twill'd made of Dog's Wool, Swingling Tow, and Posnum fur." Such whimsy reminded Americans (or at least Philly's cheeky bachelors) that Christmas, despite its moral complexities, could still be a source of childlike joy.

The Christmas tree's leap from superfrostical folk custom to global icon came in 1848, when *The Illustrated London News* published an engraving of Queen Victoria and Prince Albert standing beside a spruce imported from Albert's native Germany. To readers, the sight of a tree indoors illuminated by whale-oil candles and festooned with dolls, gilded eggshells, and candied fruit seemed revolutionary—even shocking. Victoria's influence transcended borders, and whatever she embraced, the public followed suit. The influential American magazine *Godey's Lady's Book* reprinted the image with the simple caption "The Christmas Tree." The engraving burst upon the pages of holiday books and newspapers, fostering widespread adoption by ordinary Americans of this distinctly upper-class German symbol of goodwill.

In 1870, President Ulysses S. Grant declared Christmas a federal holiday, not merely as a nod to tradition but as part of a calculated attempt to reunite a fractured nation still wounded from Civil War divisions. Grant believed shared celebrations could help heal the lingering bitterness between North and South.

Eight years later, around the time Thomas Edison introduced

residential electric lighting, the *New-York Tribune* published an article headlined MARK CARR'S LUCKY SPECULATION IN CONIFERS. Carr, an unemployed sawmill worker from upstate New York, loaded a cart with freshly cut spruces and pines, hitched it to his best ox, and traveled to New York City. There he paid one dollar for a permit to set up America's first Christmas tree stand at Washington Market, a bustling produce bazaar near today's World Trade Center. After Carr explained that the trees were intended to be stood upright and decorated indoors, his entire stock sold out in a single day. Carr's commercial success launched the seasonal tree-stand tradition, normalizing the buying and selling of evergreens as decorations, not just as lumber or firewood.

By 1889, lobbyists from burgeoning electric utilities, aiming to showcase electricity's potential to transform American life, persuaded President Benjamin Harrison to erect the first Christmas tree at the White House. Wary of the fire risks posed by electric bulbs, Harrison opted to light the eastern hemlock (*Tsuga canadensis*) with traditional whale-oil candles instead. A decade later, President William McKinley refused to publicly display a Christmas tree at all, relegating the evergreen to the maid's kitchen after letters to the *Chicago Daily Tribune* criticized the decidedly German custom as "un-American." In 1912, amid growing tensions in Europe on the eve of World War I, President William Howard Taft's mischievous children placed a spruce in the Blue Room parlor while their parents were away inspecting the nearly completed Panama Canal. Such presidential endorsements, whether intentional or accidental, cemented the Christmas tree's place in American culture. Soon, electrified evergreens began glowing everywhere, illuminating White House parlors, department store windows, and public schools across the country.

As Christmas trees became firmly intertwined with American celebrations, they continued to reflect society's darker divisions and carried forward unresolved tensions dating back to the era of slavery.

Proponents of the Lost Cause—the myth that the Civil War was fought over states' rights rather than slavery—attempted to rewrite history using romanticized fictions of enslaved people celebrating Christmas. Between its founding in 1894 and the stock market crash of 1929, the United Daughters of the Confederacy published scores of memoirs, novels, and short stories depicting enslaved families dancing joyfully around Christmas trees and accepting gifts from their enslavers—scenes deliberately crafted to obscure slavery's brutal realities. The Lost Cause's aggressive rebranding reached its disturbing apex in 1946, when Walt Disney embraced similar themes of holiday cheer among enslaved people in his most controversial and embarrassing film, *Song of the South.*

Renowned American abolitionist and social reformer Frederick Douglass sharply contradicted these misleading narratives in his 1845 autobiography. Describing his own experiences, Douglass explained how enslavers cynically manipulated Christmas to curb resistance, allowing brief periods of rest as a calculated pressure release—a bitterly ironic gift wrapped in cruelty. "These holidays serve as conductors, or safety-valves, to carry off the rebellious spirit of enslaved humanity," Douglass wrote.

In *Yuletide in Dixie*, historian Robert May underscores how fleeting and fraught Christmas celebrations were for enslaved people. Typically, the holiday respite was overshadowed by anxiety about imminent reassignment. On January 1, many enslaved individuals were rented out or traded to new slaveholders, subjecting them to fresh uncertainties and cruelties. Even when feasts and presents appeared under Christmas trees, the threat of violence and upheaval was never far away. Historical records detail chilling cases of slaveholders gathering their families around Christmas trees and exchanging human beings as gifts.

Louis Hughes, who escaped enslavement, recounted such a moment in his memoir, *Thirty Years a Slave*: "At length, after a long and wearisome journey, we reached Pontotoc, McGee's home, on Christmas eve. Boss took me into the house and into the sitting room, where all the family were assembled, and presented me as a Christmas gift to the madam, his wife. . . . I was gradually growing to the idea that it was useless to cry, and I tried hard to overcome my feelings."

Acknowledging these uncomfortable histories is essential for understanding the profound ironies woven throughout America's relationship with evergreens. For as long as Christmas trees have symbolized holiday cheer, they have also signaled who matters, who does not, and who gets invited to participate in seemingly innocent reindeer games.

Examples of exclusion and prejudice connected to Christmas trees abound in modern history. In segregated Orlando, Florida, in 1927, Santa Claus stopped only briefly at a Christmas tree in a Black neighborhood while en route to a Whites-only celebration. During the Great Depression, in 1932, a bauxite mining company in Arkansas delivered festive evergreens and bushel baskets of food to its White workers immediately after firing all of its Mexican employees. Nazi Germany appropriated pagan Yule log rituals to symbolize racial purity, and in 1953, New York City erected a Christmas tree in Union Square as a Cold War rebuke of Soviet communism. As late as 2015, the Ku Klux Klan disrupted a peaceful Muslim rally in Irving, Texas, simply because it coincided with the town's Christmas tree-lighting ceremony.

Even the Vatican historically refused to embrace the evergreen, dismissing Christmas trees for centuries as remnants of Protestant Germany's "Tannenbaum religion." This resistance finally ended in 1982, when Pope John Paul II introduced the first official Vatican Christmas tree in Saint Peter's Square, transforming a once contentious symbol into an accepted part of Catholic observance. John Paul II explicitly linked the evergreen's symbolism to the birth and resurrection of Christ, calling it "Christ, the tree of life," thus merging the tree's pagan origins with Christian theology. Since then, the towering Vatican Christmas tree has become a celebrated focal point of Rome's holiday rituals, affirming the evergreen's universal role as both cultural and spiritual emblem.

Those quick to decry a so-called "war on Christmas" might pause to reflect on how evergreens have long marked fault lines between inclusion and exclusion, power and prejudice. As decorated evergreens proliferated and spread into public places, deeper questions inevitably arose about their meaning. Had these trees become symbols of racial division, instruments of political ideology, or tools for cultural dominance? Did they still carry genuine religious significance, or had they evolved into something altogether different?

In 1984, the US Supreme Court addressed all these questions in the case of *Lynch v. Donnelly*, which centered on a baby crèche depicting the birth of Jesus Christ in a Pawtucket, Rhode Island, shopping district. Critics argued the Nativity scene violated the US Constitution's Establishment Clause, separating church and state. The court ruled by a narrow 5-to-4 margin that Christmas trees and other holiday decorations could be displayed on public property as long as the overall presentation remained predominantly secular. The court's majority reasoned that because baby Jesus appeared alongside reindeer, a Santa Claus house, a banner reading SEASONS GREETINGS, and, notably, a decorated evergreen, the diverse symbolism diluted any purely religious message. Christmas, the justices argued, had transcended its religious roots to assume broader cultural significance. The ruling effectively secularized Christmas trees, ensuring their continued prominence in government buildings and public venues.

But this decision has proved controversial. In California in 2021, a woman sued the Carmel Unified School District after officials refused her request to display a six-foot inflatable menorah alongside the school's Christmas tree. School administrators argued that while the menorah explicitly symbolized Judaism for Jewish people, an evergreen decorated with red balls was purely secular and held no religious significance for Christians. US District Judge Beth Labson Freeman upheld the district's decision, citing *Lynch v. Donnelly*. "Although Christmas trees once carried religious connotations," she wrote, "today they typify the secular celebration of Christmas."

Judge Freeman's ruling stirred unease among religious minorities, who sensed an implicit message: Hanukkah was a foreign holiday, while Christmas was inherently American. Political blogger Mia Brett expressed this frustration in 2023. "As a Jewish woman whose mother loves putting a Christmas tree up, I would really like to say that none of this matters and just enjoy the holidays," she wrote. "Far from including everyone, these demonstrations usually just serve to show Jews that we are treated as alternative Christians and that the dominant society doesn't bother to actually learn about our traditions."

Not everyone shares these concerns. Si Spiegel, a Jewish World War II veteran and the original patent holder of the first plastic Christmas tree made in America, raised his children to embrace their Jewish heritage yet saw no conflict in enjoying a Christmas tree at home. "Do you think Christmas trees were really a religious symbol?" he asked *The New York Times* in 2021. "They were pagan symbols. My kids liked them."

By the time the Supreme Court's ruling became settled law, the Christmas tree had already transcended these contentious debates and established itself as a universal totem of the holiday season. Public venues began displaying evergreens to celebrate not only Christmas but also Hanukkah, Kwanzaa, and winter solstice festivities. People of various faiths, and of no faith at all, erected trees, drawn to their festive aesthetic. Increasingly playful versions dubbed "holiday trees" popped up in every imaginable form—made of beer kegs, lobster traps, tumbleweeds, shipping pallets, canoes, and neon-lit silhouettes. Even zhuzhed-up palm trees were appropriated into holiday symbols, most famously through Corona beer's "O Tannenpalm" advertising campaign, one of the longest-running holiday commercials in history.

Whatever form it takes or context it inhabits, the Christmas tree remains a living bridge between the sacred and secular, between ancient ritual and contemporary spectacle. Persistent anxieties about belonging and identity still linger, but the tree has outlasted myths and emerged relatively unscathed from modern debates about its meaning.

Worshipped, discarded, condemned, and reinvented, it reflects our contradictions and humanity's complicated bond with evergreens. This relationship began millions of years ago, rooted in the prehistoric forests of Archaeopteris—the ancient tree that laid the foundation for humanity's eventual rise. Ever since, its resilient descendants have reliably returned each winter as defiant bursts of green against the fading light.

Which was true, until it wasn't.

When the evergreens themselves began disappearing, an unsettling question surfaced: What would Christmas be without the tree?

3
Tragedy of the Commons

▲ ▲ ▲

LONG BEFORE EVERGREENS became holiday symbols, they provided something far more fundamental: the raw ingredients for survival. Compelled by necessity and fear alike, the first humans to cut down trees were not engaged in sacred rituals but making tools, kindling fires, and raising shelters. Civilization was born from this primal negotiation between humankind and evergreens, though the precise moment of transformation remained hidden for millennia, buried along the banks of Zambia's Kalambo River.

In 2019, beneath the thundering spray of Africa's second-highest waterfall, archaeologists brushing sand away from a seemingly ordinary stick realized it was not random driftwood. It was tapered into a sharpened point and lay beside another wood fragment that had been carved into a wedge. Deeper excavation exposed something extraordinary: a four-and-a-half-foot log resting atop a larger trunk, connected by a precise V-shaped notch. High-resolution imaging revealed chop marks, the unmistakable signature of deliberate human craftsmanship.

Uranium-series dating stunned the scientific community. The logs were 476,000 years old, crafted more than 100,000 years before *Homo sapiens* even existed. Published in *Nature* in 2023, this discovery shattered long-held assumptions about early hominins, showing them to have been far more than nomadic toolmakers. Our ancient ancestors understood timber not merely as fuel or shelter but as a resource that allowed them to master their surroundings. With these notched logs at

Kalambo Falls, humanity took its first decisive step toward reshaping the planet itself.

For as long as humans have sought shelter from the elements, they have built structures from whatever materials nature provided. Ice Age Siberian hunters piled woolly mammoth bones into sturdy huts. Bronze Age Scots carved entire villages from slabs of rock. Plains Indians stretched bison hides over pine poles to make portable teepees, and Inuit communities constructed homes from ice. Yet for most of human history, across continents and cultures, civilizations built their futures from trees. Cut, carved, and stacked, wood marked humanity's transition from nomadic to settled life.

Not all wood was created equal. Slow-growing deciduous trees like oak, maple, and mahogany yielded dense timber ideal for furniture and ships, but their heavy trunks and sprawling branches made harvesting and processing them difficult. Conifers, on the other hand, offered significant advantages. Shooting skyward in straight columns rather than twisting outward unpredictably, evergreens produce uniform timber well-suited for milling into planks and beams. But the secret to their remarkable strength and consistency lay hidden deep within their anatomy.

Conifer wood consists of billions of interlocking cells called tracheids, finer than human hairs and stacked like Lego bricks into orderly lattices. These microscopic tubes run parallel to the trunk and transport water and nutrients upward from roots to leaves. Making up 90 percent of a conifer's total mass, tracheids lend evergreen lumber its characteristic stiffness, durability, and consistent grain. Deciduous trees, by contrast, contain far fewer tracheids, relying instead on shorter, irregular fibers and vessels for structural support. This key anatomical difference results in less cellular overlap, creating wood with weaker cohesion and less predictable performance, compared to the structurally superior lumber from evergreen conifers.

Tracheids are bound together with a glue-like polymer called lignin, which renders conifer wood rigid enough to bear immense weight yet flexible enough to sway without breaking. In the absence of lignin, branches would hang like cottony threads, and trunks would sag like wilted celery. Evergreens respond to fierce winds by thickening these tracheid walls, reinforcing themselves against future storms. Friedrich Nietzsche's maxim "What does not kill me makes me stronger" is encoded within every cell.

Ancient builders intuitively grasped the exceptional properties of evergreens, incorporating their wood into some of humanity's earliest architectural achievements. Around 2550 BC, while Egyptians were building the Great Pyramid of Giza, entire pine forests fell to provide timbers for dragging 2.3 million stone blocks across desert sands. Hundreds of laborers strained against ropes attached to wooden levers and sleds, which creaked under the immense weight of the stones destined to build Khufu's eternal tomb.

By about 1500 BC, evergreen timber harvesting in Greece had evolved into a specialized professional occupation. Homer vividly depicted this flourishing industry in the *Iliad* and the *Odyssey*, describing Athenian forests that "toppled with huge crashing," leaving behind "boundless timber piled." In the fifth century BC, enormous wooden crossbeams and triangular trusses made of pine harvested from the forested hillsides surrounding Athens supported the roof of the Parthenon. Remarkably, construction receipts documenting the purchase of these pine beams still survive today—mundane accounting tablets etched in stone, preserved in museums as tangible proof of a thriving lumber industry more than twenty-five hundred years old.

Yet among all ancient construction timbers, none held greater mystique or commanded more reverence than the legendary cedars of Lebanon (*Cedrus libani*). Sporting wide, multistemmed trunks and stately branches, these cedars produced a timber reputedly indestructible, resistant to decay, and so stable that it weathered centuries with minimal distortion. Their richly scented resin permeated the wood, offering a fragrance so potent it lingered long after the trees had fallen, filling ancient palaces, temples, and fortresses with an aroma synonymous with luxury.

Lebanese cedar was cherished by early civilizations of the Fertile Crescent and the Nile River Valley not merely for its prestige and aroma, but for its unparalleled utility. References to cedar reach back to humanity's earliest known written script. In the *Epic of Gilgamesh*, composed in the third millennium BC, the Sumerian hero journeys to Cedar Mountain, a forest said to have been planted at the dawn of creation. He cuts down its tallest trees to craft the mighty gates of the city of Nippur, whose cedar panels stood with imposing mythical significance.

Centuries later, around 1000 BC, Phoenician shipwrights recognized Lebanese cedar's unique water-resistant properties and used it to construct vessels capable of conquering the Mediterranean's unpredictable seas. Bas-reliefs carved into the walls of Assyrian king Sargon II's palace (circa 700 BC), now housed at the Musée du Louvre in Paris, vividly depict cedar logs being meticulously cut, rolled, and loaded onto waiting ships in bustling Phoenician harbors.

Cedar's sacred allure reached its zenith during the biblical era, becoming woven into the fabric of spiritual life. The Bible itself references "cedar"—a term encompassing both the tree itself and the wood—a total of 103 times. Specifically, the famed Lebanese cedar (Hebrew *arez*, Greek *kedros*) appears in the Bible seventy times, more frequently than any other tree. Indeed, no other plant, animal, or object earns more mentions in scripture, with the generic term "fish" ranking a distant second, appearing just forty-nine times.

King Hiram of Tyre, whose ranks reportedly swelled with a hundred thousand axmen, famously dispatched skilled carpenters to Jerusalem, helping his ally Solomon construct a majestic temple out of aromatic cedar. Its roof beams transformed the Temple of Solomon into a place where worshippers breathed perfumed air, a sensory reminder of divine grace. In ancient Lebanon's mighty cedars, humanity discovered a timber that could outlast empires, an aromatic wood whose fragrance defied time, and a symbol whose significance extended far beyond its practical uses.

During the Roman Empire, between 500 BC and AD 476, woodworking expanded into an unprecedented industrial enterprise. Elite Romans constructed lavish homes from cedar and cypress, residences known as *domus*—the root of our modern words "domicile" and "domestic." Architects relied extensively on evergreen lumber for scaffolding, brick molds, water pipes, and mechanical cranes capable of hoisting stone slabs to great heights, enabling ever more ambitious building projects. Rome's woodworking prowess peaked in AD 421, when builders constructed the city of Venice atop an upside-down evergreen forest: millions of pine and spruce pilings driven deep into tidal silt. These submerged timbers, hidden beneath Venetian palazzos and bell towers, remain marvels of human ingenuity more than sixteen hundred years later.

Romans also had a special affinity for sandarac (*Tetraclinis articulata*), a rare juniper native to Morocco's Atlas Mountains. Highly coveted for its rich, cinnamon hue and swirling wood grain, it became the wood of choice for Rome's wealthiest collectors, who filled their homes with exquisite sandarac furniture. Stoic philosopher Seneca famously succumbed to sandarac's allure, amassing an extraordinary collection of five hundred dining tables.

Behind this extravagant consumption lay a bustling, complex lumber trade, complete with taxation, logistics, and supply-chain management, which mirrors contemporary commercial operations. Roman lumber merchants hauled whole pine logs on ox-drawn carts through narrow, cobbled streets. Historical records recount traffic jams so severe that heavy timber wagons shook buildings to their foundations and occasionally toppled onto bustling streets, tragic incidents in which pedestrians were sometimes crushed beneath runaway timbers.

The scale and intensity of Rome's timber industry reinforces how society's reliance on evergreens for empire building and urban development has remained remarkably unchanged across millennia. Today's construction managers would readily sympathize with the ancients' logistical challenges: sourcing the ideal timber, ensuring timely deliveries, and preventing costly accidents.

The fall of Rome in AD 476 marked not the end of but a transition of its timber traditions into the architecture of the Middle Ages and the Renaissance. Roman timber-framing techniques became foundational throughout Europe. In Jerusalem, the eighth-century Al-Aqsa Mosque rose atop towering cedar pillars. In England, builders completed Westminster Hall in 1399, its iconic hammer-beam roof creating the largest clear-span medieval ceiling ever built—an architectural triumph built from wood and weighing approximately 660 tons. Such construction elevated master carpenters into esteemed figures, guardians of a craft valued more highly than gold.

Yet no forest lasts forever, nor does the timber harvested from it, unless it is meticulously preserved. Wood survives through the centuries only in exceptional circumstances: submerged and sealed off from air, like the waterlogged pilings supporting Venice; or kept entirely dry, as with the cypress lid of Egyptian king Tutankhamen's coffin, still bearing the immense weight of his sarcophagus after thirty-two hundred years. Without these extremes, exposed wood inevitably succumbs to moisture, air, and decay, leaving behind few clues of humanity's ancient mastery of intricate woodworking techniques.

This vulnerability created history's greatest blind spot. Ancient civilization evokes marble columns and granite ruins, yet most homes, temples, and cities were built predominantly from evergreen timber. Stone structures survived precisely because of their durability, while wooden structures decomposed gradually into soil and memory. The Parthenon's enormous pine roof beams rotted away long ago, and even Rome's sophisticated cranes depended on timbers now lost forever, leaving behind a deceptive impression of a world built solely from stone. But behind every surviving monument lies countless vanished trees, testifying to an invisible empire of wood.

Societies burned through evergreens not only for construction projects but also, literally, as fuel. Wood, it turned out, could be shaped into a home or ignited to cook food, warm bodies, and harden tools. At an eight-hundred-thousand-year-old campsite in northern Israel, archaeologists uncovered charred elephant bones and Paleolithic flint tools indicating the remains of a cooked meal. Evidence suggests that mastering fire—specifically harnessing the combustive power of wood—was humanity's first technological leap, setting the stage for both the rise of civilizations and ecological tragedies to come.

Early humans discovered by trial and error that evergreen conifers burn hotter and longer than other trees and are uniquely prone to ignition. Their branches, needles, and cones naturally exude a sticky, resinous sap that can easily catch a spark and erupt into flame. Recognizing this flammability, humans learned not just to manage these volatile materials, but to depend upon them.

The availability and importance of firewood for early humans varied widely, shaped by climate and local forests. Except for Arctic societies like the Inuit, who burned whale blubber for heat, every community depended on a consistent source of firewood for survival. In colder climates, access to reliable wood supplies meant the difference between life and death.

Yet wood's transformative potential extended far beyond merely providing warmth. Nearly every significant human achievement has hinged upon unlocking the energy stored within evergreens. The Sumerians of Mesopotamia burned trees in pottery kilns to make bricks for city walls and temples. The Egyptians razed forests to fuel metal smelters that extracted copper and tin from ore, which then fused into fire-hardened bronze weapons that armed the legions shaping empires. Metal plows turned soil for crops; wagon wheels carried goods to markets. Wood fires boiled seawater to extract salt, which enabled food preservation and stable human settlements. Charcoal—the black carbon residue from smoldered wood—formed gunpowder, revolutionizing warfare and pyrotechnics, while charcoal's humble presence in cave drawings marked humanity's first artistic expressions. Glass emerged from fire's crucible in sand melted at 3133°F, a process aided by mixing in potash—the alkaline ash of burned trees—creating "forest glass." Potash also became essential in soap production.

Most never see the flames making their lives possible. The soap just cleans; the pencil just writes; the fireworks just explode. But without wood, none of these would exist.

No society burned through wood as voraciously as Rome. At its peak, the empire stretched across three continents, powered by industries that ran on firewood. Metallurgy, salt production, glassmaking, and construction used two Mediterranean conifers predominantly: umbrella pine (*Pinus pinea*) and Aleppo pine (*Pinus halepensis*). When war broke out, as it often did, entire forests vanished into the furnaces that forged metal daggers and scabbards, shields and swords. A single bronze or iron forge could burn through five thousand pine logs daily, and across four centuries of war, Roman metalsmiths alone consumed over seven hundred million trees.

Rome's decadent bathhouses were the most extravagant wood consumers. Beneath vast steam rooms, polished marble floors concealed thousands of clay pipes forming complex subterranean hypocaust heating systems. To generate steam, bathhouse attendants splashed water onto the hot surfaces above wood-fueled furnaces, enveloping bathers in luxurious warmth. Archaeologists have uncovered elaborate Roman bath complexes boasting ornate mosaic floors, vivid frescoes depicting mythological scenes, intricate plumbing, and dedicated rooms where lumbermen tirelessly chopped and stacked pine logs, stoking roaring furnaces around the clock. The Baths of Caracalla, accommodating sixteen hundred bathers simultaneously, consumed ten tons of firewood every day. Vast fields of charred debris now mark bathhouse sites, stark evidence of Rome's insatiable appetite for evergreen timber.

Romans visited baths not merely to cleanse but to socialize, conduct business, and engage in philosophical discourse, leisurely transitioning from *tepidaria* (warm rooms) to *caldaria* (hot steam chambers) and *frigidaria* (cold plunge pools). Until Christianity's rise, Romans openly embraced communal nudity in baths, viewing physical exposure as natural, dignified, and symbolic of a civilized society unburdened by shame. Bathing rituals became so central to Roman identity that emperors participated alongside common citizens, each bringing unique eccentricities to their ablutions. Emperor Hadrian famously conducted political negotiations while lounging naked in steam-filled chambers. More extravagantly, Emperor Commodus insisted on bathing eight times a day in heated pools filled with water perfumed by cedar oils.

For centuries, the relentless consumption fueling imperial decadence carried no visible cost. Humanity's footprint seemed light. Trees grew back; the land healed. But every flame that advanced human society carried a hidden cost: a withdrawal from an ecological balance that would someday demand repayment.

By the second century AD, Romans had stripped the Apennine Mountains bare, and as local wood supplies dwindled, Caesar's legions plundered distant territories. The towering pines of Gaul (modern France), Britain's countryside, northern Africa, and Germania's Black Forest all fell to Roman axes. When the Western Roman Empire collapsed in the fifth century, southern Europe was mostly deforested, its vibrant landscapes reduced to barren hills and eroded soils.

This triggered humanity's first energy crisis. Industries withered without fuel. Metal production slowed. Bathhouses cooled. Rome had burned through the very resource that built it, crippling its infrastructure. Timber shortages, compounded by endless warfare, sealed Rome's demise. Beneath the common narrative that Rome fell due to political collapse lies a simpler truth: They ran out of evergreens.

Forest destruction did not pause with Rome's downfall; it accelerated. One civilization toppled, and another rose, repeating the mistake, cutting deeper into the land. By the time Emperor Hadrian laid waste to Jerusalem and erected cedar structures atop its ruins—including the Temple of Venus—the mighty cedar forests of Lebanon were nearly obliterated. Hadrian declared the remaining trees imperial property, erecting boundary markers to protect them, but it was too late. Mature cedars fell while feral goats devoured saplings, ensuring that the forests would never fully recover. Only twelve fragmented groves survived, preserved largely by religious decree as sacred burial grounds, representing less than 1 percent of their original range.

Roman philosophers started noticing the ecological costs of such unbridled consumption. The poet Lucretius warned of the consequences of greed in *De rerum natura* ("On the Nature of Things"). Seneca, despite his collection of five hundred exquisite sandarac dining tables, emphasized the importance of self-discipline and contentment, reflecting on how swiftly human ambition could destroy nature's gifts. In his tragedy *Hercules Oetaeus*, he wrote:

> *When all his sorrowing friends began to fell*
> *The trees on Oeta's slopes, beneath one hand . . .*
> *With deadly stroke attacked the towering pine . . .*
> *It shook its rocky crag, and with a crash*
> *Whelmed all the lesser forest in its fall.*

Seneca's lament echoed timeless tensions between human progress and nature's limits, a conflict first documented by Aristotle. He witnessed firsthand how the dense groves of old-growth Atlas cedars (*Cedrus atlantica*) that once blanketed Greece's hillsides were systematically cut down for firewood, shipbuilding, and farmland. Stripped of the trees whose roots held topsoil in place, catastrophic erosion soon followed. Fertile soil washed into the sea, leaving behind the stark white hills and limestone cliffs now synonymous with Greece.

Aristotle articulated this phenomenon as "the tragedy of the commons," observing: "Men pay most attention to what is their own; they care less for what is common." Forests—owned by none, used by all, regulated by none—inevitably faced depletion. Without defined boundaries or enforced limits, shared resources encouraged short-term overuse by individuals who took more than their fair share, ultimately exhausting the resource altogether. Economists later illustrated this concept through a parable of shepherds who were given unrestricted access to communal pastures (the "commons"); each increased their flocks until overgrazing destroyed the land's fertility.

Without limits, the commons collapses. Hence the tragedy.

From the first notched logs at Kalambo Falls to the denuded hills of Greece and Rome, to the last cedars of Lebanon, evergreens fueled civilization's rise, then succumbed to its ambition. History offers an instruction: What is neglected from the past will surely be ignored in the present.

4

Timbers of Resistance

▲ ▲ ▲

TWELVE CENTURIES AGO, a determined group of Ancestral Puebloans—the Chacoan peoples—scrambled over rocky ridges high above the San Juan Basin in what is now New Mexico. They had come on a monumental mission in search of a particular treasure: trees with strong, arrow-straight trunks capable of spanning the roofs of their "great houses," multistoried dwellings with hundreds of rooms and underground ceremonial chambers known as kivas. Not just any trees would do.

Along the way, they passed through groves of pinyon pines (*Pinus edulis*), whose rich pine nuts provided vital sustenance. But their journey through these scrubby woodlands was not merely a hunt for snacks. The crooked pinyons, with their twisting limbs and multistemmed trunks, were unsuitable for the ambitious project that had brought them there.

As the Chacoans ascended farther above the San Juan Basin, the trail crested a ridge and descended toward a sheltered ravine. Here stood exactly what they sought: ponderosa pines (*Pinus ponderosa*), rising like sentinels, their thick trunks perfectly vertical, clad in broad plates of reddish-orange bark, branches clustered between 90 and 130 feet above like dancers frozen mid-motion. From this secluded stand, they would select trees worthy of their greatest buildings—timbers that would honor the heritage of their ancestors.

They wielded stone axes, hammering—rather than slicing— through thick trunks. Needles and cones shivered with each blow.

Sweat stung their eyes, their callused hands swelling and blistering, curling stiffly around rough axe handles. At dawn, workers awoke in the chill mountain air, breaking their fast with fire-roasted corn cakes and juniper tea, bracing themselves for another day's work. They murmured prayers to ancestral spirits, seeking strength from past generations who had toiled in these same forests. Day after brutal day, their souls seized upon the dutiful pounding of stone against wood. Bruised and bloodied in this state of severe toil, the wildness of the world bound the Chacoans to the forest, forged by pain beneath the searing high-elevation sun.

Weeks passed before a tree finally weakened enough to tip over. The trunk struck the ground with a resonant crash that echoed through surrounding peaks, releasing a sudden explosion of scent: sharp resin and earthy sap mingling with dust. They gathered around its base, reverently placing their palms on the grain, feeling the sun's warmth on its now-exposed heartwood, oozing resin. Puebloan elders today recount how the ponderosas embodied ancestral spirits, a sacred belief that transformed these building materials into living vessels. They hacked off limbs to shape the trunk into construction timbers, roughly fifteen feet long and ten inches in diameter, and then they faced the hardest part.

For centuries, archaeologists puzzled over how the Chacoans transported pine logs to Chaco Canyon, now part of a UNESCO World Heritage site containing the remains of twelve magnificent great houses. Between 1986 and 2015, teams at the University of Arizona's Laboratory of Tree-Ring Research scrutinized microscopic tree growth patterns and isotopic signatures of wood samples, concluding that the timber originated from isolated mountain peaks between forty and sixty miles away in all directions. Yet the Chacoans had no horses, oxen, or wheeled vehicles to ease the burden of wood transport, and the region's navigable rivers flowed away from their construction sites. The mystery persisted. Only recently, guided by the oral traditions of Indigenous knowledge keepers, has it begun to make sense.

Skilled artisans, often Chacoan elders with expertise in weaving passed down through generations, braided pine roots and yucca leaves into durable straps called tumplines. Each strap, woven tightly so it could be attached to both ends of a heavy load to distribute the weight evenly, took days to perfect. Once finished, workers wrapped the tumplines around their foreheads, allowing them to haul heavy logs suspended horizontally behind them like oversized backpacks. Sports physiology studies conducted at the University of Colorado Boulder confirmed that human necks could indeed sustain carrying the heavy beams with a tumpline over the arduous four-day, sixty-mile trek.

Tumplines also explained the existence of peculiar thirty-foot-wide ancient roadways leading into Chaco Canyon, stretched straight across rugged terrain, often visible only from high vantage points. Built in alignment with celestial bodies, these ceremonial avenues served not only the logistical needs of loggers walking with their precious neck cargo, but also spiritual support, guiding pilgrims and timber bearers alike toward the sacred heart of Chacoan civilization.

As the workers trudged mile after unforgiving mile beneath a scorching sun, their tumplines chafed raw grooves into their foreheads, sweat mixing with dust, each step an exercise in sheer willpower. On rocky ascents, their sandals wore thin, their feet battered by stone shards and thorns hidden in the sand. Arriving at the building site, the Chacoans expertly notched each log, lifted it into position, and chinked gaps with sandstone blocks and mud mortar. Then they retraced their route into the mountains, resuming the everlasting cycle of pounding and tipping and lifting great vicious tree trunks.

The thought of schlepping whole trees through the desert using only one's forehead might seem daunting. For an entire civilization to perform this grueling task nearly a quarter million times over three centuries without wheels or draft animals seems downright impossible. But schlep they did, erecting the largest structures built by human hands anywhere in pre-Hispanic North America. To put this into perspective, Chacoans moved hundreds of thousands of logs over a cumulative distance equivalent to marching from New York to Chicago on foot, surpassing even the brute force that erected Stonehenge or assembled Egypt's monumental pyramids.

But the Chacoans did not enjoy the fruits of their labor for long. Within a century of completing their monumental great houses and underground kivas, they abandoned Chaco Canyon. Nearby mountains stood barren, their slopes stripped of mature pines and left vulnerable to erosion. Without roots to stabilize the soil, torrential rains carved deep gullies, forever altering the landscape. This demographic pattern of initial expansion, intense resource use, then collapse and abandonment would echo through Puebloan history and, indeed, foreshadow the broader story of America.

Centuries after Chaco Canyon was abandoned, Puebloans dispersed into new territories, carrying their woodworking traditions and reverence for evergreens across the Desert Southwest. For the Hopi peoples, descendants of the Ancestral Puebloans who settled in Arizona's Mogollon Rim, evergreens became integral to their ceremonial life. In elaborate rituals, Hopi dancers painted their faces with mineral pigments and wore collars, armbands, and belts woven from evergreen boughs. They chanted prayers that reverberated across desert mesas, their footsteps mimicking the pattering rain they implored spirits to summon. Farther east, the Tewa peoples of New Mexico held evergreens sacred in their spiritual cosmology, believing humankind first arrived on Earth by climbing the branches of an evergreen rising from a lake. Mountain Pima communities covered the graves of their deceased loved ones with mats woven from evergreen branches, affirming the trees' symbolic role bridging life, death, and renewal.

But when conquistador Francisco Vázquez de Coronado marched into Puebloan lands in 1540 seeking mythical cities of gold, his soldiers brutally imposed Spanish authority, initiating a devastating era of violence and spiritual desecration. The Puebloans' special knack for building wooden structures caught the attention of a Franciscan friar named Marcos de Niza. He helped the conquistadores institute a feudal system that forced Native Americans living in the area to convert to Christianity, and enslaved Puebloan carpenters to build Catholic missions across the West. According to the Spaniards' quasi-religious doctrine, read aloud by conquistadores to the Puebloan carpenters, failing to submit to their demands would have dire consequences: "We will take you and your wives and children and make them slaves, and as such we will sell them, and will dispose of you . . . and will do to you all the harm and evil we can."

Hopi leader Edmund Nequatewa later described how his ancestors were forced to cut ponderosa pines on Arizona's San Francisco Peaks for mission construction: "These beams were cut and put into shape roughly and were then left till the next year when they had dried out. Beams of that size were hard to carry and the first times they tried to carry these beams on their backs . . . if any gave out on the way he was simply left to die. There was great suffering. Some died for lack of food and water, while others developed scabs and sores on their bodies." Those who survived bore permanent physical and emotional scars, reminders of a forced labor that severed Puebloans from their lands and denied their fundamental dignity.

At Mission San Bernardo de Aguatubi, established in Hopi territory in 1629, enslaved craftsmen constructed the buildings directly over a sacred Chacoan ceremonial kiva abandoned five centuries earlier. Working under the watchful eyes of Spanish authorities, and without lumber mills at their disposal, Puebloan carpenters shaped ponderosa logs with crude stone and metal tools. Hand-stripped bark lay scattered, its sharp fragrance mixing with sweat and fear, as the peeled trunks lent the mission its distinctive, rough-hewn appearance. Secretly, many artisans etched traditional motifs into the hidden ends of beams in acts of silent defiance.

Catholic priests enforced rigid social hierarchies under a thin veneer of piety, using torture to punish Puebloans who deviated from strict European customs, especially targeting Lhamana, gender-nonconforming or third-gender individuals in Zuni Pueblo tradition. Any resistance triggered merciless Spanish reprisals: Puebloan leaders were jailed or hanged; children were torn from families; and entire communities were violently subjugated. At Acoma Pueblo, after an uprising in 1599, Spanish soldiers enslaved six hundred Puebloan women and children, condemning all Acoma men over the age of twenty-five to foot amputation, a gruesome act intended to prevent further rebellion. Such atrocities seeded deep resentment and despair that lingered throughout the region for generations.

At the heart of this suffering lay the true grift of Spanish colonialism: its relentless culture of domination, which exacted a devastating toll on Native American women. The Franciscan Fathers running the missions observed rampant illness among their female captives, particularly incurable venereal diseases like gonorrhea and syphilis. Far from embodying Christian ideals of community and peace, the missions became hubs of suffering and contagion, fueled by the systematic abuse inflicted upon Puebloan women as Spaniards marauded across the West searching for gold and timber. Ultimately, more than thirty-seven thousand Native Americans imprisoned within Catholic missions perished from disease, starvation, and violence.

But the Indigenous spirit of the Desert Southwest persisted, unbroken. On August 10, 1680, Puebloan carpenters surged through Santa Fe's narrow streets, armed with bows, clubs, and ceremonial firebrands. They toppled statues and pulled wooden crosses from church steeples. By setting fire to the very timber-framed missions their ancestors had been forced to build, they reclaimed both their land and their dignity. Amid flames illuminating the desert sky, they expelled the Spanish occupiers, killing nearly four hundred, including twenty-one missionaries, and driving the remaining two thousand south into Mexico.

Though Spanish rule would eventually return twelve years later, the Pueblo Revolt profoundly reshaped the cultural and religious landscape of the Southwest, permanently weakening colonial control and securing the survival of Pueblo culture and heritage. The memory of burned missions became a powerful symbol of Indigenous resistance.

"If they had lost, we would not be here. That is what was at stake in 1680," said Jon Ghahate, cultural educator at Albuquerque's Indian Pueblo Cultural Center. "I would not be here, and the languages of our ancestors would not be here."

Led by descendants of the original Chacoan carpenters who carried ponderosa pines across desert sands, the Pueblo Revolt stands as America's first revolutionary war, predating the war for independence from Britain by a century.

5

American Eden

▲ ▲ ▲

ONE OF AMERICA'S most cherished founding myths revolves around the 102 British Calvinists who anchored the *Mayflower* off Plymouth, Massachusetts, in 1620. History textbooks portray these Pilgrims as religious separatists fleeing persecution, but religion was not their sole motivation, nor perhaps even their main one. Their transatlantic voyage was backed by London-based merchants known as the Plymouth Company, including Sir Ferdinando Gorges, an ambitious nobleman whose fortunes had declined as England's timber reserves dwindled. For these investors, America's untouched forests represented a chance to replenish England's exhausted supplies of wood—and for Gorges, personally, an opportunity to restore his lost wealth. In exchange for funding the Pilgrims' journey, Gorges and his partners expected repayment in the form of valuable commodities—chief among them wood—to be shipped back to Britain and continental Europe.

The Pilgrims, in other words, were timber merchants.

By the early 1600s, England had become a tastefully appointed ecological wasteland: rich in culture and innovation, but desperately poor in trees. Centuries of unchecked logging had reduced its thickly forested countryside to barren moors and sheep pastures. King James I issued royal proclamations limiting timber harvests and banning the export of shipbuilding wood, but these regulations came too late. England's forests had already dwindled past recovery, echoing the fate of Rome and Greece before it.

Early European explorers—Christopher Columbus, John Cabot,

and others—had returned from the New World with tales of forests blanketing North America's eastern seaboard. England, envisioning endless timber waiting just beyond the horizon, overlooked one crucial detail: The Pilgrims knew next to nothing about logging. Most had never swung an axe before. They were farmers and textile weavers, not woodsmen. Extracting timber from a distant continent presented an altogether formidable challenge—but they would have to learn. After all, the Plymouth Company had already paid for the expedition.

Armed with Bibles, muskets, swords, and axes, the Pilgrims set sail toward an unfamiliar wilderness to chop wood for King James I. At the time, it seemed like a perfectly reasonable way to stock a supply chain. What could possibly go wrong?

The *Mayflower* spent ten grueling weeks at sea until it reached the Americas. In cold mid-December, the colonists emerged from cramped quarters below deck into a world whose scale and wildness surpassed imagination. Before coming within sight of land, they smelled it: woodsmoke, salted cod drying on racks, the pungent musk of deer hides mixing with Atlantic brine. Sailing across Provincetown Harbor, they glimpsed smoke spiraling from campfires, fishing nets draped over boulders, and purple quahog shells piled along the shore.

Far from an untouched Eden, the Pilgrims encountered a landscape meticulously shaped by Indigenous peoples for more than ten thousand years. From Long Island to Maine and from the Hudson River to Cape Cod, around 150,000 Native Americans had built thriving societies in the Northeast, supported by fertile agriculture and abundant seafood. They harvested trees sustainably for shelter and tools, and carefully managed growth cycles to yield abundant nuts, berries, and medicinal plants.

Central to this stewardship were the sacred eastern white pines (*Pinus strobus*), which the Algonquian-speaking tribes native to Long Island, New York, call *wompi koowa*. Native Americans harvest white pine resin to seal their dugout canoes (mishoons) by making precise incisions in the trees' bark during certain moon phases, ensuring minimal damage and allowing trees to heal. Indigenous mishoon builders carefully fell these trees and use small fires to gently hollow out the trunks, gradually burning and scraping away charred layers. This method allows craftsmen to shape canoes without splitting or weakening the wood, a technique embodying deep ecological understanding and spiritual reverence. The sweet inner bark of white pine also preserves dried fruits and strengthens ropes twisted from deer tendons. Pine needles, shoots, and the dense pith of pine knots—wood hardened where branches once joined the trunk—provide medicinal remedies for coughs, fevers, and infections. To Native Americans, the forest is neither commodity nor wilderness, but a carefully tended sanctuary.

Yet when the Pilgrims dropped anchor, they saw something entirely different: opportunity. Cloud-piercing evergreens towered above the beachhead, some trees five feet thick and two hundred feet tall, surpassing Europe's grandest cathedrals. Their tiered branches sagged under pine cones, ripe and heavy with seed. Here stood enough timber to build ships, homes, and cookfires for generations—a veritable wellspring of prosperity. Exhausted but hopeful, the Pilgrims interpreted this abundance not as the result of Indigenous stewardship, but as raw material for British profit. God, they believed, had delivered them to a promised land, which would soon host the biggest timber-cutting scheme the world had ever known.

Their enthusiasm soon collided with brutal reality. Winter weather arrived mercilessly, and towering pines without a single branch on their first eighty feet of trunk offered scant protection from icy winds whipping off the Atlantic. Sleeping aboard the crowded, damp *Mayflower*, colonists ventured ashore periodically to hunt game and gather firewood. Their first hesitant swings of the axe felt clumsy and uncertain as their arms trembled under the unfamiliar weight of iron axeheads imported from Sheffield. The handles, hastily carved from local maple, blistered their soft palms raw.

Yet gradually, their motions grew confident and fluid, transforming them from timid farmers and weavers into lumbermen by grim necessity. Each strike became not merely a swinging of arms but a full-body movement: raised blade, muscles tensed, knees flexed, hips pivoting, as the greedy axe bit deep into pine flesh, sweet sap spurting, and white wood chips flying.

As sleet and snow pelted down, rivers flooded, pigs ran away, wolves devoured their chickens, and flatbread slices grew thinner and thinner, colonists chopped trees, stacked logs to dry, and lit older piles aflame, desperately hoping that salvation lay in their newfound skill.

It was not enough. By spring, half the colonists had died from hypothermia, frostbite, and pneumonia, their malnourished bodies unable to withstand the bone-chilling cold.

Survivors fashioned crude huts from branches and cleared patches of forest, hoping to find fertile soil beneath fallen pines. Instead, they found barren sand inhospitable to crops. Without fishhooks to catch the boiling schools of cod offshore, they teetered near starvation. While desperately searching for better land, they stumbled upon a clearing just inland from the rocky coast, sheltered behind a line of Atlantic white cedars (*Chamaecyparis thyoides*) that once protected a thriving village. Empty wigwams stood abandoned, their bark coverings sagging and weather-beaten, their cooking hearths filled with ashes from fires long since extinguished. At the village's edge, partially woven baskets and discarded fishing nets lay tangled among dried grasses, hinting at lives interrupted suddenly and violently. Then came the discovery whose grim reality shattered the Pilgrims' illusions of virgin wilderness. Emaciated bodies littered the campsite, frozen where they fell, victims of smallpox brought by earlier European explorers. The disease had devastated the Patuxet, a band of the Wampanoag confederation, killing almost all except one man: Tisquantum, whom history remembers as Squanto. The abandoned Patuxet site would become Plymouth Colony, the first British settlement in America.

Astonished by Squanto's fluency in English, the Pilgrims saw in him a "special instrument sent of God" tasked with helping them survive. Squanto's deft language skills, however, came from a bitter past. In 1614, English explorer Thomas Hunt kidnapped him along with several other Native Americans, selling them into slavery in Málaga, Spain. Local Catholic friars intervened, freed Squanto, and arranged for him to travel to London, where he worked for John Slaney, a merchant involved in colonial ventures. There, Squanto learned English and European customs.

In 1619, he returned to North America with a trading expedition, only to discover that the Patuxet had been wiped out by disease, leaving him utterly alone: a man without a tribe. Ironically, the language forced upon him during his years abroad became the Pilgrims' lifeline in the New World. He taught the colonists how to fish, plant crops, and utilize the resources of the land, but his alliance with them was forged from necessity, not benevolence.

The colonists' religious worldview centered on dominion over nature rather than harmony with it, a belief reinforced by sermons delivered loudly beneath the rafters of newly built churches, where pastors quoted scripture exhorting them to "subdue the earth." These sermons reminded parishioners that prosperity was God's unmistakable sign of favor, making each fallen tree divine proof of their righteous mission. Forests existed to enrich individuals and, by extension, entire nations, a perspective explicitly justified by Judeo-Christian scripture. Genesis 9:2 of the New King James Bible declares, "And the fear of you and the dread of you shall be on every beast of the earth, on every bird of the air, on all that move on the earth, and on all the fish of the sea. They are given into your hand." Thomas Jefferson would later echo this mandate: "Those who labour in the earth are the chosen people of God."

And labor they did. The Pilgrims wasted little time converting forests into profits. From then on, the vanguard of America's mercantile spirit would be armed with an axe.

Subsequent shiploads of timber merchants arrived and took to the forest with haste. Every morning, the rhythmic thud of axes rang along the New England coast as pines fell, their shadows replaced by sunlight that scorched the delicate ferns beneath. Corn and beans sprouted between the stumps. Colonists widened ancient Native footpaths into crude wagon roads snaking through the underbrush, forming an ever-expanding network of trails that became frozen and impassable in winter, muddy quagmires in spring, and dusty tracks in the heat of summer. Moose retreated inland, taking with them the Wampanoags' traditional source of meat and hide. Wolves, deprived of prey, turned desperate and dangerous, leading colonists to organize hunts that offered bounties for pelts. Rivers once shaded by lush canopies became exposed and choked with silt, suffocating native trout spawning grounds and drastically reducing fish populations that Indigenous communities had depended on for generations.

Mohegan chief Miantonomi plainly described this devastation: "These English having gotten our land . . . with axes fell the trees." But there always seemed to be more trees ahead, an endless supply yielding to axes from dawn to dusk.

What began as tentative coexistence between the colonists and Indigenous tribes quickly devolved into conflict. Frustrated Native Americans "repented of their hospitality, and were inclined to dispossess their new neighbours, as the only way of restoring the country to its pristine state, and of recovering their usual mode of subsistence," wrote the eighteenth-century clergyman and historian Jeremy Belknap. In one ambush, warriors left the mutilated body of an Englishman as a warning. Colonists responded by placing bounties on Indigenous scalps. Peace proved impossible.

One winter morning in 1637, colonial militia encircled a fortified Mashantucket Pequot village near the Mystic River, setting fire to a newly constructed timber palisade fortress. Seven hundred Pequots perished, unable to escape the inferno, and their bodies were buried hastily beneath freshly cleared pine groves that settlers would soon convert into profitable farmland. The Pequot defeat accelerated colonial expansion, radiating inland from Providence and Boston over the ensuing decades. English settlements devoured vast stretches of Native territories along Long Island Sound, the scorched earth marking the merciless advance of axes and muskets.

A radically new concept in land ownership aided this expansion. Historically, Native Americans maintained forests collectively, passing land-use rights through oral agreements and generational stewardship. But English colonists imposed alien ideas drawn from Roman property law, granting ownership to those who cleared and cultivated the land. Deeds were scrawled hastily on parchment in barely legible handwriting, signed with European names Native Americans could scarcely pronounce, let alone contest. Ownership now signified more than territory; it conveyed political agency, autonomy, and personhood.

As Native populations were increasingly displaced or dead, forests once held communally became private property—assets easily bought, sold, or stolen. Indigenous peoples wishing to reclaim ancestral land faced impossible requirements: purchasing with English currency; deciphering deeds written in a foreign tongue; and paying annual taxes to avoid losing their homelands yet again to colonial authorities. An elder of the Massachusett tribe named Cutshamekin struggled bitterly in colonial courtrooms, attempting to defend his ancestral lands, only to face documents stacked against him within a legal system hostile to his traditions.

A new power dynamic came into play. Whoever controlled Indigenous land also controlled the rights to extract natural resources from it. At the heart of America's burgeoning free-market ethos lay the rapid privatization and clearing of forests, boosting timber into America's first major industry, with an economic impact surpassing the lucrative fur trade.

Early sawmills harnessed the power of waterwheels to drive crank mechanisms that moved straight saw blades in an up-and-down motion. Lumber production thus depended entirely on freely flowing rivers, especially during spring floods. New England's forests echoed with the clang of sawmills running day and night, whenever rivers provided enough current to drive the heavy machinery. Mills became hubs of frenetic activity, the air thick with flying sawdust and shouted instructions as workers maneuvered heavy logs toward pounding blades. Farmers who had once struggled to maintain subsistence amassed considerable fortunes from riverside sawmills.

Around this rapidly expanding timber economy, a vibrant society quickly took shape. Coastal towns like Portsmouth and Salem developed into bustling ports, their harbors crowded with ships loading freshly sawn lumber bound for Europe and the Caribbean. More settlers arrived weekly, drawn by the promise of new lives financed by timber. Colonial villages became polyglot communities, anchored by noisy saloons, churches, general stores, and mills processing grain and wool. And to build this growing infrastructure, towering white pines fell by the thousands, their lush canopies reduced to stark fields of stumps along the Atlantic Seaboard.

Despite being thrust into a wilderness that bore no resemblance to their European homelands, colonists stubbornly imposed familiar patterns and customs upon the landscape. They constructed homes in the English postmedieval style: rectangular saltboxes framed with sturdy wood beams, externally clad in split clapboards, and internally finished with thin woven sticks, known as wattle. They frequently set vast acreages ablaze, clearing the way for tightly clustered farmsteads behind palisade fences, their muddy yards crowded with chickens and domestic hogs.

Constructing a single saltbox home required the sacrifice of up to eighty mature pines, their trunks arduously split and shaped by carpenters wielding specialized tools called froes and adzes. Building a fireplace demanded a remarkable effort to set a stone hearth wide enough to accommodate a fully yoked team of oxen dragging logs indoors and rolling them into place. These timbers, known as backlogs, burned around the clock during long winters while colonists huddled beneath coarse wool blankets, coughing in rooms thick with woodsmoke. Their fireplaces were notoriously inefficient, sending most of the heat billowing uselessly up the chimney. Such lavish excess required colonists to build an ever-present stockpile of logs waiting to be rolled inside and burned, giving rise to the modern definition of "backlog": a surplus supply.

This profligate consumption of wood, driven by dreams of material prosperity, accelerated forest clearing and fueled the colonists' relentless pursuit of more homes, more livestock, more land, more guns—more of everything. Settlements grew unabated, pushing deeper inland, propelled by an unstoppable zeal for progress.

Timber framed the nation, both figuratively and literally, bank-rolling America's rapid expansion at devastating human cost. As the demand for export commodities soared, colonists forced Native Americans into sawmills and logging camps, leading the region's transition toward race-based slavery. Just as the Spanish had enslaved Puebloans in New Mexico, British colonists enslaved Indigenous peoples, compelling them to fell pines and saw them into planks, supervised by cruel overseers who whipped those unable to meet impossible quotas. Under constant pressure, the laborers' backbreaking work yielded barely one hundred board feet per day, far too little to satisfy the colonists' ambitions. (One board foot measures twelve inches by twelve inches by one inch.)

Yale University historian Ned Blackhawk underscores this tragedy in *The Rediscovery of America*, noting that during the sixteenth and seventeenth centuries, "the Indian slave trade . . . exceeded the African trade," with colonists enslaving almost one million Native Americans. This exploitation quickly expanded into a global commerce in captive woodcutters, as colonial merchants trafficked enslaved Natives across triangular routes connecting Boston, London, and the Caribbean. Many captives found themselves shipped to remote logging camps in Belize and Jamaica, chained in damp, suffocating ship holds before enduring grueling labor hauling logs through insect-ridden jungles. Those distant colonies had no sawmills of their own and shipped whole logs back to Boston, where still more enslaved Indigenous workers processed them into export-ready lumber.

As timber flowed from forests to sawmills and onward to distant markets, colonists confronted an unexpected crisis. They had commodities in abundance, but British currency was dangerously scarce. England, itself running low on precious metals, refused to supply sufficient gold and silver coinage to its colonies. Merchants struggled to conduct daily transactions and resorted to barter—exchanging lumber, grain, livestock, and bullets—but cumbersome trade quickly reached its limits. Forest products, despite their great value, proved unwieldy as payments. Logs and sawn planks, difficult to transport and impossible to divide, could neither settle debts nor pay taxes. An economy thriving on timber needed a more stable tender.

Beneath this practical dilemma lay a deeper truth. New World prosperity rested explicitly upon Indigenous dispossession, forced labor, and ecological plunder. America's original Eden, its vast forests commodified and consumed, was rapidly being transformed by profit-seeking entrepreneurs—even as the colonists struggled to translate that newfound commercial success into tangible currency.

In 1652, desperate colonial leaders took matters into their own hands. Acting without British permission, Massachusetts minted a coin designed to be defiant, durable, and unmistakably American. Boston silversmiths melted down stolen Spanish silver, often seized from ships or traded illicitly, and hammered out coins by hand, stamping each one with the silhouette of their conquest: the eastern white pine. America's first currency, the pine tree shilling, bore not an emblem of a distant monarchy but the tree whose fate symbolized the young nation's relentless pursuit of wealth, and the devastating human price paid to achieve it.

6

The Pine Tree Riot

▲ ▲ ▲

EMPIRES RARELY BEGIN with riots; usually, they end with them. America, however, not one for tradition, ignited its revolution not over tea or taxes, as it is commonly told, but over pine trees. It seems absurd that evergreens could unravel a global imperial order, but that is precisely what happened. The seeds of a rebellion that toppled kings and rewrote history were planted deep in the soil of New England's forests.

By the early 1700s, as Britain and France competed fiercely for North American dominance, naval strength became paramount. Just as a bomb today is useless without the aircraft deploying it, eighteenth-century power depended on two-hundred-ton warships capable of carrying one hundred iron cannons. The ships were historically outfitted with sailing masts harvested from perfectly straight hundred-foot-tall European evergreens, but trees of that size no longer grew on British soil. Their predicament carried its own circular logic: Wooden ships existed primarily to procure more wood. Britain needed vessels large enough to ferry pine logs from distant forests back to London shipyards, supplying the construction of even bigger ships—and so on, in a maddening, unsustainable cycle.

Other empires had been quicker to secure their mast-building supplies. Portugal owned shipyards in Brazil. Spain harvested forests in the Balkans. France still had the pines of Vallée d'Aspe, while Italy zealously protected the Dolomites' red spruce (*Picea excelsa*), prized for musical instruments as well as shipbuilding. England traditionally relied on Riga fir, also known as Scotch pine (*Pinus sylvestris*), harvested along the Baltic Sea from dense and shadowy stands of evergreens rising like fortress walls from snow-clad shores. But those forests lay beyond narrow straits controlled by Denmark and Sweden. Fully aware of England's dependence on their ports, these Scandinavian kingdoms levied exorbitant taxes, often forcing British naval officers into humiliating dockside negotiations, watched closely by stern-faced customs agents.

Sweden's internationally renowned iron industry also depended on the Baltic region's abundant Riga firs, which provided inexpensive fuel for smelting furnaces that Atlantic Europe could not match. The rapid expansion of the French, Spanish, and British navies required casting tens of thousands of iron cannons, driving Swedish woodsmen deeper into their forests until most of their timber supply had gone up in smoke. As European forests dwindled to less than 10 percent of their original range and exporting large timber through the Baltic straits grew more expensive, Britain was left with few viable options to maintain its naval fleet.

The British Empire's global aspirations thus rested entirely on the most mundane of foundations: America's eastern white pines.

New England's forests stretched for hundreds of miles in an unbroken emerald expanse sustaining countless mature specimens reaching between 220 and 270 feet, their trunks wrapped in thick, furrowed bark. They were as durable as Riga fir but a quarter of its weight, their wood finely grained and arrow-straight, ideal for the rigors of ocean navigation. The British Crown confidently assumed American colonists would eagerly supply mast timber, but colonial loggers were decidedly unimpressed. Much to the Royal Navy's chagrin, American sawmill operators saw little financial value in mast-pine harvesting. Asked why they shunned their duty to serve the mother country, colonists pointed to the nightmarish drudgery involved.

Before touching a mast-worthy pine, loggers first had to spend weeks cutting paths through dense underbrush from the nearest navigable waterway, sleeping in crude lean-tos beneath canvas tarps that barely kept out the chilling rain. Mosquitoes swarmed relentlessly, loggers' fires smoked and sputtered, and their meals consisted of stale bread and salted pork eaten hastily in the gloom before sunrise. Each new path they carved grew progressively longer with every harvested tree, forcing crews deeper into the wilderness, far from help should accidents strike.

Once they had hacked their way to a mast-worthy pine, they felled dozens of smaller trees around it to lay down as cushioning so it would not shatter upon impact. Taking down a mast pine required precise calculations and a healthy dose of dumb luck. Loggers would carefully circle the trunk, scrutinizing the slope of the ground, studying the direction of prevailing winds, and judging the tree's natural lean. They tapped into what some call "bush sense," an extrasensory gut intuition for how the trees would behave in any given situation. After determining the correct course of action, they chopped deep notches into the trunk, the axe blades dulling quickly against resinous wood. Every swing of the axe sent nerve-wracking vibrations up the trunk, shaking loose dead branches that could, and often did, plummet with lethal force.

When available, loggers wielded European crosscut saws to slice through trunks over the course of several exhausting days, lubricating the blades with whale oil carried in whiskey bottles hanging from their belts. Miscalculations could cause a tree to unexpectedly lean backward, trapping a saw blade mid-cut and leaving loggers nervously coaxing a forty-ton trunk that teetered unpredictably on its stump, threatening death with every gust of wind. The test for which loggers had the best skills of their trade, and which didn't, was a terminal pass-fail.

Even after successfully felling a mast pine without getting injured— an achievement itself—trees frequently toppled sideways from intended landing zones, rendering them impossible to move, or dragging neighboring trees down in chaotic entanglements of branches and roots. Ninety percent of felled trees proved useless due to hidden decay and were bucked into manageable pieces and sent to nearby sawmills. The rare perfect mast tree, perhaps one in every ten, required a hitch of forty oxen to drag it to port. Any pine more than twenty miles from shore proved effectively unreachable. When a logger finally delivered a mast, it fetched perhaps $55 (about $2,750 today), paltry compensation for such difficult work.

Instead of supplying mast timber to Britain, colonial lumber merchants chose the far more lucrative and convenient business of milling pines into construction planks—work often performed by enslaved Native Americans. Every pine cut into boards, however, meant one less mast available for warships, a reality that drove British officials into fits of frustration. Colonists' reluctance was not merely economic stubbornness, but a clear sign their allegiance to the Crown was fraying.

Responding to what Britain saw as egregious violations of Crown property rights, Parliament passed the 1711 Act for the Preservation of White and Other Pine Trees in America for the Masting of Her Majesty's Navy, forbidding colonists from cutting any white pine over twenty-four inches wide on public land. British wardens, stern men in scarlet coats emblazoned with royal insignia, fanned out across New England's rugged countryside, wielding specialized hatchets to carve upward-pointing broad arrows into the trunks of the largest, most desirable pines. The distinctive arrow symbol—blunt, arrogant, a constant reminder of distant royal authority—appeared overnight on trees, warning colonial lumbermen of severe penalties should they dare fell Crown timber.

Colonists were livid. They undermined British mast demands by cutting marked trees anyway, then hiding evidence of their misdeeds by milling logs into twenty-three-inch planks, one inch below the legal limit. Homes built before the Revolution proudly featured twenty-three-inch pine plank flooring, a potent statement of rebellion. British spies infiltrated colonial logging camps hoping to persecute the worst offenders, but colonists evaded British capture by slipping quietly into moonlit forests, their faces darkened with soot, masquerading as Native Americans. In the silver glow, they worked swiftly and quietly to cut forbidden pines, muffling axe blows with cloth wrappings. By morning, the woods stood eerily silent, the marked trees having vanished without a trace, leaving bewildered British wardens to puzzle over phantom woodsmen.

An exasperated British Parliament tightened restrictions further, banning the cutting of all white pines, regardless of size or location. The royal surveyor of the woods dispatched a sheriff to Weare, New Hampshire, to arrest sawmill operators accused of illegally harvesting mast trees. The stage was set for confrontation.

On April 14, 1772, the sheriff, weary from his journey, stopped for a drink at Weare's Pine Tree Tavern, a modest clapboard inn with sagging roof shingles and a weather-beaten sign depicting an evergreen tree. Inside, a fireplace crackled loudly beneath exposed pine beams, its walls alive with boisterous talk of local grievances. The sheriff sipped his ale, oblivious to the resentment simmering among the tavern patrons and the storm gathering around him.

A defiant posse of disgruntled lumberjacks stormed in, their faces still smeared with soot, led by the brash Ebenezer Mudgett, a forty-six-year-old liquor merchant and sawmill owner known locally for his outspoken contempt of British authority. A veteran of countless disputes with Crown officials, Mudgett and his friends dragged the hapless sheriff outside, held him upside down by his ankles, and whipped him mercilessly with pine boughs—an ironic punishment delivered with savage glee. The humiliated sheriff fled on his horse past a gauntlet of jeering townsfolk. News of the brawl quickly spread throughout New England, galvanizing anti-Crown sentiment.

The Pine Tree Riot directly inspired the more famous Boston Tea Party the following year, and soon white pine became synonymous with colonial defiance. George Washington's newly commissioned Continental Navy proudly unfurled the Pine Tree Flag atop colonial warships' masts carved from the very trees Britain had claimed as its own. As this new flag snapped in the wind, it symbolized more than naval rebellion; it was a declaration of sovereignty.

In 1775, the Continental Congress banned pine mast exports entirely, forcing the Royal Navy to improvise by cobbling together masts from undersized pines bound with iron hoops. When the British fleet sailed from the West Indies toward Virginia in 1781 to relieve General Charles Cornwallis, besieged at Yorktown, violent Atlantic storms splintered the hastily improvised masts. Captains shouted orders lost in roaring winds and sailors struggled desperately to rig temporary repairs, but the masts disintegrated under relentless assault, sending sails, ropes, and wood crashing into heaving seas. British ships were left helplessly adrift. With rescue impossible, Cornwallis surrendered at Yorktown in the decisive, final battle of America's War of Independence. A revolution rooted in New England's forests had succeeded, forever linking the idea of freedom to a nation of wild evergreens.

7

Babbitt's Big Idea

▲ ▲ ▲

IN THE EARLY nineteenth century, America's population was bursting at the seams, crowding into narrow brick row houses and wooden tenements whose poorly ventilated rooms became breeding grounds for cholera and tuberculosis. By the end of Thomas Jefferson's second presidential term, the young republic forged by revolution had grown to seven million, pushing eastern port cities to their limits. In rapid succession, the founding of the New York Stock Exchange in 1817 and the completion of the Erie Canal in 1825 heralded a wave of urban growth, stoking new industrial demands for timber.

Forests of pitch pine (*Pinus rigida*), with its thickly fissured bark and its branches heavy with resin, fueled bakery ovens, brick kilns, and sprawling saltworks whose boiling rooms stretched three smoky miles along the canal. Workers in these saltworks labored under oppressive conditions, their faces streaked with soot, breathing air so dense with smoke and salt vapor that their clothing and hair became coated in a gritty crust by day's end. Timber not burned as fuel went directly into sawmills, where fully a third came out the other end as sawdust—a waste dumped along riverbanks and set ablaze in towering, smoldering heaps. Sawdust fires routinely ignited floating logjams, setting entire rivers aflame for weeks. Waterways became so choked by smoke and embers that ships navigated through them only by lighthouse beams, ironically powered by still more burning pine. The antiquated methods of axe-wielding loggers inherited from Pilgrim ancestors fell hopelessly short against the needs of a nation on the move, and bottlenecks in timber supply threatened to stall America's expansionist ambitions.

The solution to this crisis came from an unlikely source: a woman named Tabitha Babbitt, whose creative genius blossomed within the Shaker community in Harvard, Massachusetts. Officially the United Society of Believers in Christ's Second Appearing, the Shakers were an egalitarian society famed not only for its communal living and meticulously handcrafted furniture, but also its revolutionary approach to gender roles. In a world dominated by male inventors and industrialists, Shaker beliefs provided rare opportunities for women like Babbitt to flourish as skilled artisans and innovators, free from societal constraints.

Her transformative insight came around 1813 when she noticed sawyers struggling with the inefficient two-man pit-saw method used widely in sawmills. In this arrangement, one worker stood atop a log positioned above a deep pit dug in the ground, while another stood in the hole beneath, both alternately pushing and pulling an eight-foot-long whipsaw lengthwise through the log. But the teeth of the whipsaw blade were specifically angled to cut only on the downward stroke, rendering half of the workers' efforts futile. Babbitt envisioned a radical improvement: a circular blade mounted on a rotating axle, spinning continuously in one direction and enabling uninterrupted cutting as logs were steadily fed into it.

According to Shaker tradition, Babbitt demonstrated her concept by attaching a notched tin disk onto her spinning wheel, inviting skeptical community elders and mill workers to watch as she pushed a length of wood against its spinning edge. Witnesses reportedly marveled as the rudimentary blade smoothly parted the wood, its deafening whir signaling the birth of a revolution in lumber processing.

Remarkably, Babbitt chose not to patent her groundbreaking invention. Committed deeply to Shaker principles that discouraged personal profit from communal innovations, she believed ideas should serve the greater good. Though Shaker records consistently credited Babbitt's ingenuity, no physical prototype survived history, and she received little public acknowledgment during her lifetime.

Almost overnight, Babbitt's spinning toothed disks transformed quaint, riverside sawmills from modest two-man operations processing three thousand board feet daily into roaring factories staffed by hundreds cutting over forty thousand board feet. The 1840 US Census documented more than thirty-one thousand wood-processing enterprises operating east of the Mississippi, roughly one for every five hundred Americans. The sheer number of sawmills churned out a torrent of lumber to drive industrial growth as inventors and craftsmen turned wood into everything imaginable—clocks, furniture, shoes, sewer pipes, trestle bridges—setting the stage for an imminent building boom.

Two centuries after the *Mayflower*'s landing, New England's eastern white pine forests had grown thin, nudging lumbermen west across the Appalachians into New York, Pennsylvania, and Ohio. With each new region logged, land speculators pressed westward still, drawn to the reportedly inexhaustible North Woods of the Lake States: Michigan, Wisconsin, and Minnesota. Here, virgin pines stood thick and wild, dark sentinels in dense groves so tangled they blocked out the sun, creating twilight at noon.

Civilization charged headlong into the North Woods, propelled by a roving impulse inherent to the new American spirit. Eastern investors claimed the largest tracts of land and unleashed a ruthless, no-holds-barred assault pitting steel saw blades against evergreens. The woods rang with the metallic hum of sawmills as speculators raced to convert standing trees into stacks of cash.

Thick swarms of mosquitoes descended upon fresh human targets, their incessant buzzing driving loggers into fits of distraction and despair. Men desperately slathered their faces and hands with bear

grease to ward off the marauding insects, but welts still swelled their
eyes shut and left their skin raw and bloodied by evening. Despite
the mosquitoes' valiant defense, settlers continued their push into the
wildwood, harnessing the power of Tabitha Babbitt's big idea to build
bridges and canals around river rapids, timber slides bypassing water-
falls, and horse-drawn sleds that crisscrossed snowy terrain.

Lake States lumber production exploded just as the nation urgently
needed new housing for its growing immigrant population, but traditional English-style homes with heavy timber frames simply could
not be built fast enough. After years of experimentation, American
carpenters seized on a radical new construction technique called balloon framing, which replaced bulky beams with lightweight stan-
dardized boards fastened with metal nails. Balloon framing relied on
dimensional lumber, uniform two-by-four studs that could be mass-
produced cheaply, allowing builders to erect structures in a fraction of
the time using significantly less material. Factories also supplied pre-
fabricated doors and ready-to-install windows, reducing on-site labor
and transforming home building from artisanal craft into industrial
enterprise.

Initially mocked as flimsy contraptions that could float away in
the wind, balloon-framed houses became one of the most important
technological developments in human history. Embracing simplicity
over complexity, balloon framing exemplified a phenomenon known
as "worse is better," in which practicality and ease of execution tri-
umphed over traditional standards of aesthetic refinement. Unlike
post-and-beam homes, which relied on expensive, complicated joinery
that could be executed only by highly trained craftsmen, balloon framing succeeded precisely because virtually anyone capable of swinging a
hammer could learn how to build a house.

Balloon-framed houses built with dimensional lumber and nails
were marginally less sturdy than European timber frames, but they
rapidly democratized homeownership in America. Previously, homes

were symbols of elite craftsmanship and inherited wealth, but balloon framing enabled newly arrived immigrant families and frontier farmers to construct affordable dwellings. Across prairies and mountain valleys, clusters of hastily assembled homes symbolized not only American innovation but also a profound shift toward egalitarian ideals.

This new building technique spread nationwide along with the circular saw blade, binding American homes irrevocably to the nation's evergreen timberlands. Cedar shingles covered roofs; spruce lath anchored plaster walls; knotty pine paneling lined interiors. As balloon framing reshaped communities, the widespread adoption of steam power liberated sawmills from their reliance on waterways, allowing lumber towns to spring up anywhere trees stood and transforming America's landscape into a humming grid of timber enterprises. Yet with cities, factories, and farms increasingly dispersed across vast distances, efficient land transportation became paramount.

With their proud funnel-shaped smokestacks billowing thick plumes of black smoke and brightly painted drive wheels pounding steadily over steel rails, woodburning steam locomotives became instantly recognizable icons of America's nineteenth-century golden age of rail travel. The piercing whistle of these iron giants signaled their arrival and departure from stations crammed with immigrants, farmers, merchants, and the constant fragrance of pinewood smoke.

The nation's first railroad opened in 1830, initially aiming to connect the Chesapeake Bay to the Ohio River. Three decades later, rail lines stretched to the Mississippi River and eclipsed horse-drawn wagons as the primary mode of commercial transport. By the mid-nineteenth century, America's rail network spanned over thirty thousand miles, reorienting the nation's geography by shifting transportation routes from north-south river corridors to east-west railroads.

The expansion of this rail system required an immense amount of firewood. Steam locomotives stopped every twenty-five miles at bustling crossroads known as "wood-up" stations, where entrepreneurs shoveled pine into hungry fireboxes while passengers stretched their legs and vendors hawked hot coffee and baked pies. Each year, locomotives incinerated approximately ninety million cords of firewood—enough to build seven hundred thousand balloon-framed homes. (A cord measures eight feet long, four feet high, and four feet wide.)

This continuous, voracious consumption of firewood accompanied a feverish demand for dimensional lumber in Chicago and Milwaukee, where tens of thousands of immigrants arrived by train and clamored for balloon-framed houses. As lumber and settlers flowed perpetually westward, railroads accelerated the Anglo-American Settler Revolution, fundamentally reshaping the nation's economy and demography in a single generation.

The pace of growth proved particularly explosive in the Midwest, where hundreds of millions of acres of Indigenous territories were ripe for the taking. Railroad companies secured westward expansion through federal land grants to unceded Indigenous territories, which guaranteed the railroads unfettered access to forty-mile-wide corridors through tribal lands across the continent, plus a $16,000 payment for every mile of track installed on level ground (about $500,000 today). Everywhere new tracks appeared, construction booms followed. It had taken Europeans two hundred years to secure the eastern half of the continent, and now trains would enable them to conquer the West.

Thousands of agriculturally inclined immigrants crossed the Mississippi River and built settlements on the Great Plains, a grassland comparatively devoid of trees, with no landmarks, just open sky. Stretching hundreds of miles across ocean-like distances, the prairie shocked newcomers. One US Army private, after days of travel, cried out at his first glimpse of distant trees: "Be Jesus, we're in sight of land again!" This trackless enormity spared settlers the toil of clearing forested land but left them utterly reliant on lumber imported from the Lake States.

As railways connected small rural communities to urban trading hubs, the resulting boom in ranching and farming turned what many once considered a treeless wasteland into an agricultural mecca. Railways brought millions of cattle and hogs to Chicago's sprawling stockyards, themselves built from pine planks and cedar fence posts. Homesteaders in remote settlements dotting the plains of Kansas and Iowa eagerly awaited shipments of dimensional lumber arriving by rail, transforming isolated prairie outposts into thriving communities with the construction of barns, silos, fences, churches, and the iconic "little houses on the prairie." Wood that was once a regional material exclusive to the wealthy immigrant class in the East became an essential commodity affordable to even low-income families and Midwestern farmers. The railroads brought not just lumber and livestock, but also capital and credit, linking settlers in the heartland to eastern investment banks and consumer lending institutions, driving a symbiotic surge of settler colonization and financial speculation.

This whirlwind of European immigration, industrial expansion, and technological innovation accelerated timber harvests across the Lake States. Amid the rush, however, quiet warnings went largely unnoticed. William Strickland, the renowned architect who designed the first US Mint building in Philadelphia, cautioned prophetically: "What [timber] is not wanted for any present purpose is set fire to. If care not be taken it will soon be very scarce." Loggers in heavily harvested areas began murmuring uneasily about forests reduced to stark clearings, where soil dried and eroded, clogging streams and silencing birdsong in once-thriving canopies. But such concerns were drowned out by the relentless clatter of Tabitha Babbitt's saw blades and the booming whistles of steam locomotives, quieting the first voices of conservation beneath the irresistible roar of progress.

Scarcity seemed impossible in a land teeming with such visible abundance. Forty million bison roamed the Dakota grasslands. Passenger pigeons, numbering five billion strong, could block out the sun for hours over eastern cities. Fifty thousand grizzly bears stalked prey from the Sonoran Desert to the subarctic frontier. White men had not yet discovered oil in Texas when immigrant farmers carried their axes into the forests, determined to carve their destinies under the unwavering conviction that America was, indeed, a land of unlimited opportunity and self-made men.

8

Life on the Mississippi

▲ ▲ ▲

EVEN AS NORTHERN forests fell, America's gaze shifted southward toward a brewing conflict that would rip the nation apart and redraw its destiny. On the eve of the Civil War, timber barons eager to feed the engines of commerce fixed their ambitions on a vast frontier thousands of miles from the mosquito-plagued North Woods. The breakaway Confederacy, rich with untapped evergreen forests, offered the promise of an industrial revolution propelled by steamboats plying the country's longest river. Cotton plantations, sugar mills, ironworks, and sawmills sprang up along the banks of the Mississippi, each enterprise dependent on the constant flow of firewood in an economy fueled by evergreens and wrought by human bondage.

The Mississippi's steamboats were as perilous as they were powerful. With towering smokestacks swaying precariously, belching ash and cinders above crowded decks, they moved through murky waters propelled by machinery slapped together from small cast-iron parts riveted hastily in place. Without regulations, oversight, or basic mechanical training, reckless captains routinely pushed their vessels past the breaking point. Pressurized boilers overheated and burst, shattering wooden hulls into flying daggers. The resulting explosions filled the air with scalding steam, flaming debris, and human screams. These catastrophes were so frequent and infamous that river travelers nervously dubbed each voyage a "floating lottery," uncertain whether they would reach their destination alive.

Between 1811, when the first steamboat traveled the Mississippi, and 1853, boiler explosions alone killed more than seven thousand people on the river. Such tragedies continued claiming lives well into the nineteenth century, including that of Henry Clemens in 1858. His brother, Samuel Langhorne Clemens, later chronicled steamboat horrors in *Life on the Mississippi*, under the pen name Mark Twain: "The next moment four of the eight boilers exploded with a thunderous crash, and the whole forward third of the boat was hoisted toward the sky! . . . Many people were flung to considerable distances, and fell in the river."

Also among the victims of steamboat accidents was Charles Follen, the Boston abolitionist who popularized the decorated Christmas tree. Follen had been traveling aboard Cornelius Vanderbilt's luxurious steamer *Lexington* in 1840 when flames suddenly engulfed the ship. Passengers leapt frantically into the frigid waters of Long Island Sound, their cries echoing helplessly into the night as the steamship sank in the icy darkness. The disaster inspired one of the earliest lithographs from famed printmaker Currier & Ives, linking America's Christmas tree tradition to the hazardous technology powering the nation's industrial growth. Despite such risks, Americans eagerly boarded steamboats, lured by the promise of swift, comfortable passage at a time when overland travel remained tedious.

Yet convenience came with severe ecological consequences. At every steamboat's heart was its water boiler, a wood-burning furnace located on the main deck, where fires produced steam to drive pistons and power the paddle wheel. At their peak in the mid-1800s, some riverboats burned seventy-five cords of firewood per day, an inefficient clip even by the smoke-belching standards of the era.

Steamboat engineers preferred firewood from longleaf pines (*Pinus palustris*), which ignited easily, burned intensely, and built up a quick head of steam. These giant evergreen conifers once flourished in sandy soils along a two-hundred-mile-wide band blanketing the southeastern and Gulf coastlines. Their tall, slender trunks stood crowned with tufts of green needles spaced widely enough for sunlight to filter down and nourish vibrant meadows below, rich with wild orchids, pitcher plants, and wire grass. Adapted uniquely to periodic lightning-induced wildfires, longleaf pine ecosystems thrived through cycles of destruction and renewal. Covering ninety-two million acres, these forests helped the nineteenth-century South become a global economic powerhouse known for more than plantation-grown cotton.

Historians have often mischaracterized the Mississippi Delta's pine belt as virgin forest, unscathed by millennia of Indigenous occupation and America's first two centuries of unbridled expansion. Some assumed the cotton-based plantation economy and warm climate preserved these woods, since fewer trees were cut for domestic heating than in the North, and most of the region's rivers flowed away from the northeastern cities where firewood could be sold. Yet by the time Thomas Jefferson finalized the Louisiana Purchase in 1803, nearly a century of intensive logging had already thinned these forests. Until the development of sugar as a cash crop around 1800, French and Spanish colonists had long shipped planks, shingles, and barrel staves to French Caribbean islands like Martinique and Guadeloupe, whose economies depended on imported lumber. Over seven hundred thousand enslaved Africans formed the vanguard of this early timber extraction in the Lower Mississippi Valley, unknowingly laying the groundwork for greater destruction when steamboats arrived a half century later.

As word of the steamboat boom spread, tens of thousands of mostly Scottish and Irish immigrants flocked to the Mississippi, hoping to strike it rich selling firewood to passing steamboats. Calling themselves woodhawks, they roamed the river in crudely built flatboats stacked precariously high with freshly split pine logs. At night, their campfires dotted the riverbanks like constellations. These rough, solitary men who had fled failed farms or oppressive debts spent their evenings singing melancholy ballads, drinking cheap corn whiskey, and boasting about fortunes made and lost along the ever-hungry Mississippi. With their products always in demand, and the land and trees they exploited costing them nothing, woodhawks accumulated wealth at great cost to the region's forests and people alike.

To supply steamboats with firewood, enslaved lumbermen felled pines by axe, rolling logs down muddy riverbanks onto rafts headed downstream, where additional crews hacked them into precise lengths to fit the steamboat fireboxes. A crew of eight lumbermen working from sunrise to sunset could clear an acre of dense pine forest every day, an area that typically contained thirty to fifty mature pines. As steamboat traffic intensified and demand at wood stations surged, these logging crews were forced into an exhausting cycle of chopping, rolling, and rafting. Naturalist William Bartram described encountering enslaved lumbermen "comparatively of a gigantic stature, fat and muscular, mounted on the massive timber logs, the regular heavy strokes of their gleaming axes re-echo[ing] in the deep forests."

Once the tallest trees were felled, enslaved women and children moved through the newly opened clearings, hacking away remaining brush with hoes and hatchets. At nightfall, under calmer winds, they gathered the woody debris into piles and burned them slowly, fertilizing the soil with potassium-rich ashes but leaving behind landscapes stripped of their ancient grandeur.

Although most lumbermen were enslaved Africans, freedmen who had fled a life of cotton sharecropping often joined their ranks, as did maroons who had escaped plantation bondage and sought refuge in the delta's inhospitable terrain. Displaced Native Americans, driven from ancestral territories, also found occasional work chopping firewood alongside impoverished White farmers seeking supplemental income. Together they endured grueling conditions while leeches clung to their legs and man-eating alligators patrolled the swamps. Injuries were constant and unavoidable. Hands and feet bore the brunt of the daily trauma, crushed under falling logs, mangled by chains and ropes, or bitten by venomous cottonmouth snakes lurking in shaded hollows. Most men worked in thin deer-hide moccasins that a wayward axe could cut through like wet cardboard. Their crude axes, blunted and poorly forged, would glance unpredictably off wood, turning routine tasks into a game of Russian roulette. Occasionally, an axe-head would fly off its handle mid-swing, sail through the air, and shear off part of another man's skull. Yet dangers and exertion from cutting the trees down were only half the battle. Moving these giants to the river proved equally daunting.

Each mature longleaf pine stood more than a hundred feet tall and weighed over thirty tons, requiring hitches of at least six horses or oxen to drag it to the river's edge. Without roots to stabilize riverbanks, the mud was so deep and soft that timber carts and draft animals sank, their limbs straining helplessly. During the rainy season, tree removal became downright impossible. Lumbermen adapted by selectively harvesting smaller, younger trees, sacrificing the future productivity of the forests for immediate ease of transport.

Compared to cotton plantation slavery, the lumbermen's work was far more dangerous but provided a modicum of quasi freedom. The presence of biting flies and venomous snakes discouraged slaveholders from closely supervising the logging operations, giving enslaved men moments away from constant scrutiny. Within the secluded backwoods, they built cabins, gathered medicinal plants, and supplemented their meager diets by hunting rabbits, turtles, and opossums. Networks of social and trade relationships connected dispersed lumber camps through dimly lit clandestine trails that the lumbermen regarded with an almost mythical reverence. These hidden pathways of refuge and quiet rebellion were subtly marked by bent pine saplings and stone cairns, known only to those who walked them at night. By meeting in the woods to whisper news of freedom or to exchange precious goods smuggled from steamboats, lumbermen could create secret lifelines that defied their overseers' iron grip or simply provided space for quiet reflection.

"If any one wishes to be impressed with the soul-killing effects of slavery," observed Frederick Douglass in the autobiography of his life, "let him . . . place himself in the deep pine woods, and there let him, in silence, analyze the sounds that shall pass through the chambers of his soul."

The forest offered more than physical refuge; it was a sanctuary where voices and spirits could rise freely, if only briefly, without fear of punishment. Muffled prayers could be heard in "hush harbors," clearings hidden within the trees where enslaved people held spiritual gatherings away from White supervision, singing softly enough to evade detection.

White landowners saw these forests through a sharply different lens: as property markers, boundaries between wilderness and civilization. After their enslaved workforce had finished stripping trees from the riverbanks, they either planted cotton or sold the newly cleared parcels to farmers in an exploitative practice called "log it and flog it." This allowed landowners to extract profits twice: first from timber, then from farmland. But logging served another purpose for landowners, equally critical, by tightening their control over enslaved laborers. Fewer trees meant fewer hiding places, forcing them into open clearings where overseers patrolled on horseback.

As escaping became more difficult, some lumbermen, like Cudjo Lewis, found work slinging firewood aboard steamboats. For Lewis, every steamboat landing became a backbreaking ordeal, hurling cords of heavy pinewood up slippery gangplanks while impatient steamboat captains barked commands from the upper decks. Despite the punishing nature of the work, it offered Lewis rare moments of respite and camaraderie. During brief intervals between loadings, he exchanged news and rumors with crewmen who traveled up and down the river.

Decades later, Lewis described his experiences working on a steamboat to anthropologist Zora Neale Hurston, vividly recalling all twenty stops that the ship made: "Oh Lor! I workee so hard! Every landing, you unnerstan me, I tote wood on de boat." Born Oluale Kossola, Lewis had been kidnapped and illegally transported to America aboard the schooner *Clotilda* in 1860, the last known slave ship to import African captives. "Sometime de bluff it so high we got to chunkee de wood down two three times fo' it git down where de river is. De steamboat didn't used to burnee de coal. It burnee de wood an' it usee so muchee wood!"

Some enslaved lumbermen earned small amounts of money from occasional tips aboard the steamboats, but the luckiest seized opportunities to escape. Antebellum literature documents the punishments slaveholders inflicted upon lumbermen who fled and were recaptured: Achilles tendons sliced, limbs amputated, ears severed, and victims rolled through lumber camps in wooden barrels studded with nails. In extreme cases, slaveholders built fires to slowly roast maroons on spits using firewood the maroon himself had chopped from the forest. Castration was also common, exacting an emasculating psychological toll that stripped enslaved men of their dignity and manhood.

In the book *Rethinking Rufus*, Howard University historian Thomas Foster describes the violation of Black manhood "in other ways that are less easily spoken of (then and now), namely, the sexual exploitation of enslaved men." He underscores how slaveholders treated Black male bodies as commodities to be controlled for economic gain. Such exploitation included forced sexual pairings designed to increase the labor force, which denied enslaved people autonomy over their intimate relationships and caused deep psychological wounds to men and women alike. Enslaved men also faced direct sexual violence and rape from their White slaveholders, reflecting a societal fixation on Black men's bodies: simultaneously objects of desire, fear, and domination. Though historical records rarely mention the sexual violence specifically experienced by lumbermen in the Mississippi Delta, their isolation in remote timber camps, beyond the reach of social scrutiny, made them especially vulnerable. This unspoken abuse further reveals how deeply slavery intertwined with the extraction of timber resources from the South's great longleaf pine belt.

An entire wealthy, ruling class of White cotton farmers emerged who derived their initial fortunes from timber profits. In his 1860 proslavery novel, *The Sunny South*, written explicitly as a rebuttal to Harriet Beecher Stowe's *Uncle Tom's Cabin*, Mississippi author Joseph Holt Ingraham illustrated this linkage. He described how a twenty-one-year-old man was gifted "thirty slaves from his father, and fourteen hundred acres of wild forest land on the Mississippi" and leveraged the gift to generate substantial wealth. "He took his hands there, and commenced clearing. Thirty axes do vast execution in a wood," Ingraham wrote. "As he cleared he piled up the cloven timber into fire-wood length, and sold it to passing steamers at $2.50 a cord. The first year he took $12,000 in cash for wood alone."

This fictionalized account reflected the real-world financial strategies of many Mississippi planters who used profits from steamboat firewood operations as essential startup capital for launching cotton empires. Timber thus served not only as literal fuel but also as a financial bridge enabling settlers to scale up their cotton operations, which in turn unleashed ecological havoc. The transition from timber to cotton was seamless yet devastating. Once stripped of their protective pine forests, soils eroded, leaving behind gullied, sterile fields that drove slaveholders farther westward in search of fertile ground. These intertwined ecological and economic dynamics connected timber extraction directly to the cotton boom, forming a cycle of exploitation that degraded landscapes and amplified human suffering across the South.

Cotton plantations and steamboats represented only two dimensions of the South's burgeoning economy. Another, equally destructive, was the iron industry. Nearly 250 antebellum iron furnaces throughout Virginia, the Carolinas, Kentucky, and Tennessee depended heavily on pine charcoal as fuel. Raw firewood alone did not produce sufficient heat to smelt metal, so blacksmiths first had to convert pine logs into charcoal through a process called pyrolysis: slowly heating wood in oxygen-starved kilns to extract resin, tar, and water, leaving behind the dense carbon substrate. Every hundred pounds of pine yielded just twenty pounds of charcoal, but this concentrated fuel burned hotter and longer, making it ideal for smelting iron.

The kilns themselves were primitive and dangerous, emitting thick, acrid smoke that clung stubbornly to workers' skin and clothing. Kiln operators, mostly enslaved men, spent weeks tending these slow-burning mounds, shoveling dirt over them day and night to control airflow under the constant risk of suffocation from toxic fumes. The largest Appalachian iron furnaces required hundreds of enslaved laborers at each site to sustain charcoal production, and historians estimate around ten thousand spent their lives fueling the furnaces of the antebellum South.

Strategically built into hillsides for easy loading, some furnaces rose forty feet high, casting sinister silhouettes against the glowing night sky as enslaved men fed endless streams of iron ore, limestone, and charcoal into the fiery chambers. Each furnace consumed roughly an acre of pine forest daily to produce just two tons of iron, a ratio so low that ironmasters continually needed new expanses of woodland to maintain a steady charcoal supply. After exhausting nearby forests over a decade or two, they abandoned these smoke-blackened clearings and moved on. In their wake, newly sunlit patches of fertile soil allowed pine cones to open, releasing seeds that sprouted into dense sapling thickets.

Such a forest had grown in Virginia by 1864 at the site of the Civil War's infamous Battle of the Wilderness, where General Ulysses S. Grant's Union army first clashed with General Robert E. Lee's Confederate forces. Twenty years earlier, loggers had stripped this same land bare to fuel local iron furnaces; now millions of young pines crowded the battlefield in a disorienting maze.

As fighting erupted, pine needles muffled the soldiers' footsteps and branches clawed at their uniforms, limiting visibility to mere yards. Then artillery fire ignited the dense undergrowth, and soldiers suddenly faced a terrifying dual hellscape: enemy fire and wildfire.

Fueled by resin-rich branches and needles, the firestorm instantly engulfed dry pines in towering columns of flame that advanced on all sides and raged up to their tops. Tree limbs crackled explosively, showering embers onto soldiers below. Glowing rings of needles burned yellow, orange, then charred black, reducing trees to smoldering skeletons beneath ghostly trails of smoke. This dramatic chain reaction—known as full-tree candling—is unique to evergreen forests, where fires can leap from tree to tree like firecrackers linked by a common fuse.

By the time the battlefield fell silent, over twenty-nine thousand Union and Confederate soldiers had died, their final breaths filled with smoke and the unmistakable scent of burning Christmas trees.

Such nightmarish collisions of warfare, industry, and forest destruction altered the southern landscape and economy for generations. Once among America's richest and most diverse ecosystems, widespread deforestation in the South nearly eradicated deer, black bear, and cougars. Riparian wildlife reliant on the delicate balance between water and woodland—frogs, turtles, salamanders, and crayfish—also declined sharply. Birdlife suffered the most, with species like the ivory-billed woodpecker, the Bachman's warbler, and the Carolina parakeet driven to extinction.

Underlying this devastation was a ruthless logic shared by slavery and capitalism: extracting maximum profit without regard for human dignity or ecological balance. Across the nineteenth-century American South, and especially in the Mississippi Delta, these intertwined forces consumed both human lives and evergreen forests, etching deep scars into land and memory.

9

The Turpentine Trail

▲ ▲ ▲

IN EAST TEXAS today, families tracing their ancestry through faded deeds, marriage records, and scattered historical documents often rediscover a forgotten corridor known as the Turpentine Trail—a winding route across the antebellum South, marked tree by tree.

This trail emerged during a forestry boom that swept through the region in the mid-nineteenth century, prompted by a global crisis: the severe shortage of whale oil. Before petroleum and electricity defined modern life, whale oil was indispensable for illuminating American homes and lubricating industrial machinery. Relentless hunting along North Atlantic migratory routes, led by harpoon-wielding crews aboard ships sailing from Nantucket and New Bedford, had nearly driven humpback, sperm, and right whales to extinction. With whale populations in sharp decline, oil prices spiked dramatically in 1855, reaching $35 per gallon (equivalent to about $1,280 per gallon today). Its scarcity sparked a frantic search for affordable alternatives, leading America toward a new solution distilled from the resin of longleaf pines: turpentine.

Turpentine deepened America's intimate connection to evergreen conifers, whose distinctive scent—familiar to anyone who has smelled a Christmas tree—remains stubbornly difficult to describe. Many plants possess scents so recognizable they share a single word with their names: lavender, mint, rose. Evergreens, however, elude such simple categorization. Their aroma is primal yet comforting, evocative but intangible, captured only vaguely by the word "piney." It originates from volatile organic compounds called phytoncides, key ingredients in the highly flammable turpentine. Evergreens release phytoncides as invisible aerosols to defend themselves against airborne fungal spores that can infect wounds and digest trees from within.

Separated from resin by distillation, liquid turpentine became a valuable industrial solvent for paints, varnishes, polishes, and household disinfectants, all prized partly for their fresh pine scent. Beyond its industrial uses, nineteenth-century households also kept bottles of turpentine in medicine cabinets as questionable cure-alls, applying it externally to treat cuts and rashes, or even drinking small—and sometimes lethal—doses as a folk remedy for parasites and stomach ailments.

The US Patent Office recorded 250 separate registrations for a wide assortment of lamps, wicks, burners, and fuels during this period. Among the hopeful inventors, one stood head and shoulders above the rest: Henry Porter of Bangor, Maine. Porter patented a mixture he named camphene, which consisted of one part turpentine, four parts alcohol, and a dash of pine-scented camphor, another highly flammable essential oil derived from the camphor tree (a broad-leafed evergreen, but a laurel, not a conifer). Sold as Porter's Burning Fluid, this concoction burned brightly, emitted a pleasant aroma, and became a favored alternative to costly whale oil.

Yet camphene was dangerously volatile, infamous for household tragedies documented across newspapers of the era. Horrifying accounts detailed lamps spontaneously exploding at family dinners or on bedside tables in the dead of night. A Boston newspaper denounced camphene in 1859: "There are different kinds of lamps and lamp oil, adapted to different tastes and circumstances; and there is one at least, most abominable invention under the name of Camphene Oil, or Burning Fluid, which were better denominated a Swift and Ready Means of Destruction for Private Families." Despite regular explosions and house fires, consumer demand for camphene continued to rise in tandem with whale-oil prices. The camphene industry's future seemed bright, provided it could secure a steady supply of turpentine.

America's turpentine industry began modestly in North Carolina in 1720 but surged after Britain lifted import tariffs in 1846, allowing the state to dominate the global market and produce 96 percent of the nation's supply. Turpentine barons managed sprawling woodland holdings, often encompassing twenty thousand acres and employing hundreds of enslaved laborers across dozens of distilleries. Their dynastic fortunes rivaled and sometimes exceeded those of the South's wealthiest cotton plantations.

The process of extracting resin began each winter, when enslaved lumbermen known as dippers used specialized axes to carve deep, pocket-shaped cavities into the trunks of longleaf pines, typically six inches above the ground. Each tree hosted three or four of these collection sites to accelerate the harvest. As temperatures warmed in spring, the pines responded to their wounds by exuding copious resin, filling the freshly cut cavities. Dippers would then use long-handled metal scoops to collect roughly two quarts of the sticky amber-colored resin per tree—a gummy harvest that would cling to their tools and pull painfully at hair and flesh while they scraped out every precious drop. Working under brutal conditions, each dipper serviced around twenty-five hundred trees daily, routinely suffering injuries from slipped axes, heat exhaustion, and dizziness caused by prolonged inhalation of turpentine fumes. Dippers emptied their collections into barrels destined for central distilleries, where the resin was boiled down into highly flammable turpentine that regularly exploded, killing or disfiguring enslaved distillers.

To sustain a consistent resin flow throughout the summer, dippers would revisit their trees every two weeks and carve new incisions in the bark, leaving distinctive rows of downward-pointing chevrons that were visible from a considerable distance. Traveling through North Carolina in 1843, New York poet and journalist William Cullen Bryant described seeing pine forests that had been "boxed" for turpentine extraction. Bryant was horrified by the scars left behind as broad wounds gouged into the tree trunks. "This is the work of destruction," he lamented. "It strips acre after acre of the noble trees."

Visitors like Bryant might have witnessed the damage left by dippers but rarely encountered the enslaved men themselves. They lived deep within the remote forests, sheltered in rudimentary four-foot-tall timber shanties crudely assembled from pine planks and bark that offered scant protection from the elements. At night, their resin-stiffened overalls stood upright on their own, propped like ghostly sentries beside the shanties. In summer months, the dense forest trapped stifling heat and humidity, causing rashes and sweat sores. Rain seeped through the cracks of their shanties during storms, leaving the men sleepless on damp and moldy pine-needle beds, praying in vain for the oppressive conditions to break.

White overseers occasionally patrolled these isolated camps on horseback, but distance and dense underbrush prevented close supervision, granting dippers a marginal degree of autonomy. This required a different kind of labor regime than cotton, and dippers would sometimes receive small cash bonuses for exceeding their resin harvest quotas. But once the volatile burning fluid was distilled and shipped to markets in New York City and London, profits flowed effortlessly into the pockets of White turpentine barons just as surely as sweat streamed down the labor-hunched, keloid-scarred backs of enslaved dippers.

When America's turpentine industry peaked at ninety million gallons in 1860, few realized the industry had played a key role in relieving pressure on whale populations, helping avert their near extinction.

But the turpentine boom would soon flame out. Dippers had managed to weaken North Carolina's pine forests so extensively that trees were left vulnerable to beetle infestations and wind damage. Boxed trees—scarred and oozing flammable resin—were known to spontaneously combust on sunny days. Summer lightning strikes turned the husks of leached pines into an inferno. Dried branches, needles, and hardened resin accumulated on the forest floor, suffocating native understory species of pitcher plants, hawthorns, and daisies. Red-cockaded woodpeckers, which relied on mature longleaf pines for nesting and foraging, vanished as their habitat degraded, leaving only a lifeless tinderbox. Farmers converted these decimated forests into cotton and tobacco fields, permanently disrupting natural seedling renewal.

North Carolina's longleaf pines never recovered.

Having depleted the financial basis for their own existence, turpentine barons forcibly relocated enslaved dippers to pursue fresh, untapped forests in South Carolina. After depleting those lands, they moved on to Georgia, Florida, and eventually Alabama. Decades passed, and as these forests deteriorated, the industry pressed into Arkansas and Louisiana.

After the Civil War, politicians desperate for economic revival offered federal, state, and Indigenous timberlands in the South for sale at absurdly low prices, mere pennies an acre. Northern investors capitalized on the land rush, acquiring millions of acres of southern pine forests, hastening the ecological collapse. The Goodyear brothers, already masters of sawmill empires in New York and Pennsylvania, founded Louisiana's Great Southern Lumber Company, constructing the world's largest sawmill and cutting one million board feet of lumber from fifty acres of virgin longleaf pines *every day.*

 A parallel pencil manufacturing industry roared to life, dispatching thousands of "cedar cruisers" across southern states to follow the turpentine dippers in search of eastern red cedars (*Juniperus virginiana*) growing in mixed stands with pines. Initially, the finest pencil wood came from the heartwood cores of naturally fallen old-growth cedars that had dried in the sun. But as those became scarce, cedar cruisers turned instead to living forests, harvesting only the prized heartwood and leaving 70 percent of the remaining timber to rot. Even the grandest red cedars yielded only two hundred thousand pencils—a sobering statistic considering Americans, experiencing widespread literacy and booming commerce, bought one billion pencils annually, far outpacing nature's ability to replenish.

As cedar forests dwindled and abandoned turpentine stands burned, cedar cruisers resorted to increasingly desperate tactics. They dismantled anything they could get their hands on that happened to be made of cedar—barns, distilleries, furnaces—and convinced cotton farmers to barter their cedar fence posts for newfangled barbed-wire spools. Soon pencil makers had wiped clean the entire southern stock of red cedars, living or dead, and started making pencils out of the broadleaf deciduous tree called basswood (*Tilia americana*), which they artificially dyed and perfumed with turpentine to mimic red cedar's fragrance.

By the mid-1880s, the alarm bells began ringing loudly as the cathedral-like longleaf pine forests, once thought inexhaustible, disappeared at alarming rates. Miles of denuded and stump-filled land stretched bleakly across the South, leaving nitrogen-poor soil worth peanuts, literally, since peanuts were the only nitrogen-fixing crops that could survive on the cutover lands. From these ruined landscapes arose the American political tradition of the "stump speech," named for the tree stumps candidates running for public office stood on while delivering political orations. Fiery politicians like William Jennings Bryan and the incendiary South Carolina populist Benjamin "Pitchfork Ben" Tillman, known for his unvarnished racism, traveled from town to town along the Turpentine Trail, rallying disenchanted White farmers with impassioned speeches delivered atop stumps. These makeshift platforms came to symbolize the era's most intense economic and racial grievances, inseparable from the ecological scars left by longleaf pine extraction.

The last few turpentine barons and their descendants hung on and made it all the way to the Piney Woods of East Texas, founding the town of Turpentine and setting up a mother camp for resin collection and distillation. There, the great southern botanical vandalism finally reached its inexorable conclusion when the majestic longleaf pine forests collapsed. From an original habitat of ninety-two million acres sprawling across nine states, just twelve thousand acres survived in isolated patches: 99.9 percent of *Pinus palustris* vanished into history.

But by then, it hardly mattered. America barely paused to catch its breath before turning westward, where even mightier trees awaited discovery in the crumpled facets of California's Sierra Nevada.

10

The Discovery Tree

▲ ▲ ▲

IN THE SPRING of 1852, a rugged gold miner named Augustus Dowd, renowned for his buckskin duds and tangled blond curls, tracked a grizzly bear through the forest above the gold rush town of Murphys, California. Dowd remained keenly alert, mindful of the violent encounter two years prior when he had lost three fingers fighting a grizzly, eventually killing the beast with only his knife. This pursuit, though initially ordinary, was soon eclipsed by a discovery that dwarfed his past adventures. He abandoned his hunt out of distraction as he wandered into a grove of the biggest trees he had ever seen—trees larger, he thought, than *anyone* had ever seen. Who could possibly believe such enormity?

Dowd's fellow miners scoffed at his description, skeptical in part due to his penchant for exaggeration, but he persuaded a group of them to go see for themselves. After a full day on horseback, they reached the tree and immediately fell silent. The giant's upper branches dissolved into swirling fog overhead. Its deeply furrowed trunk descended into a pedestal base thirty feet wide, like a candle the size of a steam locomotive had melted from red desert rocks.

Dowd sent branches and cones from the mysterious giant to Albert Kellogg, a San Francisco physician who had studied plants alongside renowned naturalist John Audubon and later cofounded the California Academy of Sciences. At the time, Kellogg was completing a botanical classification begun by Austrian botanist Stephan Endlicher, who had set out to categorize all known evergreen conifer species by counting the seeds on each cone scale and organizing them numerically. Endlicher's methodical sequence began in the Old World with the Eurasian genus Agathis (pines native to the Philippines), which have between one and three seeds per scale. His sequence concluded at the upper end with Sciadopitys (umbrella pines from Japan), featuring as many as nine seeds per scale.

But a conspicuous gap existed precisely where California's massive evergreens should have fit within Endlicher's numerical sequence. Kellogg realized that Dowd's samples filled this empty slot, as they bore five to seven seeds per cone scale, exactly the number Endlicher had predicted but never confirmed in a living specimen. In honor of its position completing Endlicher's elaborate classification, Kellogg named the genus Sequoia, derived from the Latin *sequi*, meaning "to follow," underscoring how this remarkable tree had filled a critical gap in botanical knowledge.

Entire dissertations have since explored the sequoia's name origins. For decades, the popular assumption was that an Austrian botanist had named these trees after Sequoyah, a Native American from Oklahoma who invented the eighty-six symbols representing the spoken Cherokee language. The idea that the big red trees honored a "noble red man" resonated in America yet lacked credible historical evidence beyond the phonetic similarity of the two names.

Among California's ancient sequoia survivors, only two species remain: the giant sequoia (*Sequoiadendron giganteum*) and the coast redwood (*Sequoia sempervirens*, from the Latin *semper*, meaning "always," and *virens*, meaning "green"). Both conifers flourished alongside dinosaurs during the Mesozoic Era seventy million years ago, migrating up and down the continent like J. R. R. Tolkien's Ents, retreating southward during ice ages and reclaiming northern territories as glaciers receded. Though both species have distinctive rust-colored wood and are referred to interchangeably as either redwoods or sequoias, they live in different habitats. Coast redwoods thrive in humid fog belts, drawing up to a third of their water from mist rising off the frigid North Pacific, condensing on their leathery dark-green needles. Giant sequoias, by contrast, prefer higher elevations along the sunbaked western slopes of the Sierra Nevada, where melting snow-pack satisfies about half their water needs.

Sequoias defy conventional biological limits. They boast the world's tallest tree at 380 feet (equivalent to a thirty-five-story skyscraper), the widest at more than thirty feet (broad enough to block an interstate highway), and among the oldest, with one specimen documented to have lived 3,266 years in the Converse Basin Grove. Eastern white pines and longleaf pines rarely exceed five feet in girth—mere twigs compared to a sequoia. These trees do not just grow large; they grow fast, some reaching eighty feet within their first forty years, quickly overtaking smaller rivals. Yet not all sequoias become hulking giants. Those growing on high-elevation rocky slopes often remain smaller overall—stunted in both height and girth due to harsher growing conditions—while those rooted in lower-elevation river valleys with rich, moist soils achieve legendary proportions in both dimensions.

Dowd's discovery, seen in hindsight, might better be termed a *re*discovery, or the first documented sequoia sighting by a White person. A fuller narrative would acknowledge the Miwok, Washoe, and more than a dozen other Indigenous nations who inhabited and carefully managed these Sierra Nevada forests for millennia. Their societies thrived primarily in coastal regions, where protein sources required ingenuity and skill but not the extreme energetic output that deep-forest living demanded. While various Indigenous groups culti-vated fruit trees and berries in forest understories, many—including the Miwok—regarded the towering sequoias with reverence, believing ancestral spirits resided within the ancient giants. They rarely ven-tured deep into the sacred groves except for ceremonial purposes or to gather thick bark for their cone-shaped *kotcha* houses, or to hollow fallen trunks into canoes. To the Miwok and other Indigenous peoples of the Sierra, cutting down a living sequoia was not merely wasteful, but an act of profound spiritual violence.

For White settlers, however, sequoias represented irresistible wealth—trees simply too valuable to leave standing.

Despite the remote location deep in California's gold country, news of Dowd's Discovery Tree spread quickly. The *Daily Placer Times and Transcript* in San Francisco published a story describing "a cedar tree in Calaveras County 30 feet in diameter at the butt." Newspapers in New York and London heralded the tree as an unprecedented wonder of the natural world, yet skepticism flourished. The *New-York Daily Times* speculated that the tree might merely be a gold miner's elaborate fiction, "a manufactured marvel." Victorian audiences, well acquainted with P. T. Barnum's hoaxes and sideshow frauds, like the infamous Fiji Mermaid, were understandably wary. Thus arose a compelling public demand for tangible proof, prompting a frenzied quest to commodify this natural curiosity.

To prove doubters wrong, Dowd and his fellow miners stripped off the first fifty feet of the tree's bark, hauled it by horse and wagon on primitive mountain trails to Bush Street in San Francisco, and reassembled it in a spacious room with carpet, a piano, and seating for forty people. He charged visitors fifty cents admission and attracted large crowds during its two-month run, but San Franciscans still harbored doubts that so much bark could all have come from the same tree.

Enter Captain William Hanford, a California sawmill owner and president of the Union Water Company, where Dowd worked. Spotting a lucrative opportunity, Hanford devised a scheme to cut down the Discovery Tree and ship a cross section around Cape Horn for a splashy public exhibition in New York City. Like others of his era, Hanford saw nature strictly in terms of extractable wealth. So it came to pass that the first sequoia anyone cut down happened not to harvest its timber, but simply to prove it could be done.

Hanford and five men trekked to Calaveras Grove and commenced drilling hundreds of holes into the Discovery Tree's formidable bark using crude two-inch pump augers. The steel tips dulled quickly against the tree's dense fibers, forcing exhausted men to pause repeatedly and sharpen bits with whetstones and leather strops. Their efforts turned the immense trunk into a Swiss-cheese pattern of puncture wounds in hopes of weakening it enough to topple it. When the stubborn giant remained standing, they chipped away at it with chisels for several more days until finally breaking through the three-foot-thick bark. After building a platform atop the stack of debris at the tree's base, they carved out a room-sized hollow inside the trunk so they could climb up and, in a twisted crouch, swing their axes.

With each swing, they hacked deeper into the past. Hidden growth rings emerged from the trunk, rings formed when the Chacoans built their desert kivas, when smallpox devastated Squanto's tribe, when Emperor Hadrian tried in vain to preserve the cedars of Lebanon, when Hanford's own ancestors still lived as medieval serfs in England, centuries before the *Mayflower* sailed—rings laid down before King James I, Columbus, and even Aristotle were born. All this history fell unnoticed in a growing mound of redwood chips at their feet.

After twenty-two days of chiseling and chopping, on June 27, 1853, the Discovery Tree came crashing down, leaving behind a gaping hole in a forest, and a nation, that would never be the same.

Hanford arrived in New York City eager to present the sequoia cross section at the Crystal Palace, the famed glass-and-steel exhibition hall located at what is now Bryant Park, near Sixth Avenue and Forty-Second Street. He soon found himself mired in disputes with the building's new proprietor, the notoriously competitive showman P. T. Barnum. When negotiations fell apart, Hanford instead rented a venue at 596 Broadway. Determined not to be outdone, Barnum countered by announcing his own competing exhibition featuring another giant California sequoia, launching it just days before Hanford's opening. Barnum's stunt sparked public confusion, severely diminishing Hanford's attendance. Frustrated and financially depleted, Hanford closed his exhibition after only a week.

Returning to California, Hanford renamed the Discovery Tree the Discovery Stump and transformed its enormous base into a dance floor, alongside a bowling alley built atop its fallen trunk. Undeterred, he cut down an even larger sequoia and sent a cross section to London. There, at last, he found the fame he had long sought. On opening day, twenty-seven thousand visitors flocked to his display at London's Crystal Palace.

Wealthy Victorians marveled at the tree, which became a coveted status symbol. Inspired by botanists such as William Lobb, who first gathered sequoia seeds from the Sierra Nevada in 1853, British aristocrats dispatched gardeners on costly transoceanic missions to collect cones from the California groves. Sequoia saplings thrived in English soil, astonishing horticulturists by growing two feet or more each year, quickly transforming pockets of the British landscape into exotic evergreen forests.

Hanford's London success inspired more sequoia exhibitions back in America. In 1874, loggers felled a tree measuring twenty-four feet in diameter, transporting a segment of its trunk east for public viewings in Chicago and Cincinnati before ultimately selling it to magazine publisher Frank Leslie, who installed it as a garden sculpture at his Saratoga Springs estate. Sequoias also became roadside spectacles as enterprising loggers carved tunnels through living trees, creating surreal passageways wide enough for horse-drawn wagons. President William Howard Taft famously rode through the Wawona Tunnel Tree by carriage in 1909, accompanied by naturalist John Muir and politician Charles Forbes. These "drive-thru trees" became lucrative attractions, drawing crowds unaware that the damage to the trees' vascular systems would accelerate their demise.

Every time one of these ancient giants fell for pure amusement, America seemed numb to the ecological devastation. In fact, people demanded more visible proof that such immense trees existed, which encouraged even greater pillaging by none other than the US government.

Following the idea that "if people cannot visit the trees, bring the trees to the people," the US Army spent eight days in 1891 felling a sequoia from California's Big Stump Grove, using a specially crafted twenty-three-foot-long saw. Named the Mark Twain Tree in honor of the beloved author whose writings captured the awe of these giants, the tree stood 331 feet tall and was more than thirteen hundred years old when the Army shipped a cross section of its carcass to be displayed for astonished crowds at New York's American Museum of Natural History. US Interior Department officials harvested another giant sequoia, naming it after Secretary of the Interior John Noble, to amaze visitors at the World's Columbian Exposition in Chicago in 1893. Afterward, it stood for decades on the National Mall near the Smithsonian, where a park service placard proudly proclaimed: "The trunk asserted the nation's arrival as a mighty power, one whose promise and latent riches were as massive as the tree's height and bulk." Years later, custodians on the National Mall unceremoniously topped the once-grand sequoia hull with a cupola and began using it as a toolshed.

Amid this high-profile exploitation, voices of caution began to rise. Among those witnessing the destruction firsthand was John Muir, the Scottish-born naturalist who wandered through the Sierra Nevada in 1876. Horrified by logging operations denuding entire hillsides, Muir described the landscape as "bleak scenes of ruin and death." His grief intensified when he saw loggers proudly preparing giant sequoia trunks for eastern exhibitions, including the famed Daniel Webster Tree, gleefully carted off to become a sideshow attraction at Philadelphia's Centennial Exposition. In his journals, Muir bitterly noted the irony of a nation celebrating its centennial by desecrating its oldest living inhabitants.

Nevertheless, trees that had stood since the age of dinosaurs continued to be thrust into the spotlight of human ambition as both objects of wonder and targets for exploitation. The sequoias' appearance in New York and London for much-ballyhooed exhibitions accomplished more than confirming their existence to a skeptical public. They firmly established the prehistoric giants as lucrative tourist attractions and industrial commodities.

For all the attention they garnered and all the people they impressed, none would be more significant than two ambitious railroad tycoons: Leland Stanford and James Strobridge. As architects of the first transcontinental railroad, their vision reached far beyond mere circus sideshows. They saw the sequoias as inexhaustible reservoirs of raw material capable of fueling America's westward expansion and industrial dominance. It would not take long for these titans of industry to recognize the enormous fortunes waiting to be carved from the valley of the giants.

11

Too Much of a Good Thing

▲ ▲ ▲

A CROWD OF thousands gathered on Promontory Summit, Utah Territory, in 1869 to watch the once unthinkable become reality. The Union Pacific Railroad, built westward from Omaha, met the Central Pacific Railroad, which had conquered the formidable Sierra Nevada eastward from Sacramento. With one last tap of the spike maul, a cross-country journey that took six grueling months by wagon could now be comfortably completed in just eight days by train.

President Ulysses S. Grant received a telegraph dispatch from Leland Stanford, the influential founder of Stanford University and president of Central Pacific Railroad: "We have the honor to report the last rail laid, the last spike is driven. The Pacific Railroad is finished."

Telegraphs relayed the news to cities everywhere, sparking spontaneous celebrations nationwide. Chicago hosted a four-mile-long parade. The mayor of New York City ordered a one-hundred-gun salute. The bells of Independence Hall in Philadelphia rang continuously for so long that residents feared it was a fire alarm.

Shortly afterward, construction foreman James Strobridge hosted a celebratory meal in his private railcar, honoring eight Chinese men who represented the virtual army of twenty thousand others who had built the line. Railroad executives, who had earlier abandoned plans to hire Mexicans and formerly enslaved Africans, had turned instead to Chinese immigrants to form the backbone of their construction crews. Hired primarily through Chinese foremen acting as trusted intermediaries, laborers came largely from rural villages in Guangdong Province in southeastern China. They fled after their land and way of life were shattered during the Taiping Rebellion, a fourteen-year-long uprising against the Qing dynasty that killed an estimated twenty million Chinese. More than a hundred thousand migrants—mostly men—crossed the Pacific Ocean to a continent utterly foreign in language, customs, and climate. Initially skeptical of their capabilities, Central Pacific managers came to rely on the diligence and adaptability of Chinese crews, who filled nearly 90 percent of the railroad's labor needs and were by far the single largest workforce in American history to that date.

Chinese railroad workers held every position imaginable, from engineers and foremen to cooks and doctors. Stanford University historian Gordon Chang described their complex sociocultural world in his book *Ghosts of Gold Mountain*, highlighting their interactions with the wider American population, the challenging working conditions they faced, and the diverse ways they expressed their cultural identity and religious beliefs.

"Thousands more Chinese associated with them as friends and relatives, as part of the immense supply chain that provisioned them for years," Chang wrote, "and, away from the track in their off-time, as gamblers, opium smokers, prostitutes, and devout worshippers of the gods and spirits who watched over them in their perilous work."

Over five brutal years, from 1864 to 1869, Chinese laborers faced unimaginable hardships. Working through bitterly cold Sierra winters, they cleared snowbanks taller than buildings and tunneled through solid granite at elevations reaching nine thousand feet. Frequent avalanches buried entire crews alive, and dynamite blasts maimed or killed dozens at a time. They organized a rare strike in 1867, successfully demanding slightly better wages and shorter working hours. Yet an estimated one thousand perished anonymously before the railroad was completed, their exact numbers and identities lost to history—many of them buried beneath snowdrifts or scorched into oblivion by oppressive desert sun.

For their essential labor, the Chinese were rewarded with severe racial hostility. Denied the right to marry Whites or become American citizens, they faced strict limitations on employment opportunities and became targets of violent mobs, who drove them from railroad towns. Their lives, shaped by loneliness and isolation far from home, rarely entered official records, leaving only faint echoes captured later by historians. Though indispensable to America's dream of coast-to-coast travel, the "Railroad Chinese" remained nearly invisible to passengers comfortably speeding along tracks built with their blood and sacrifice.

Other speakers at James Strobridge's private railcar celebration thanked the Chinese workers but hewed closely to Stanford's grand narrative, calling the railroad a "great steel horse," poetically declaring the "iron chain" would finally "marry the Atlantic and Pacific Oceans." No less significant was the symbolism of those statements. Just four years after the Civil War ended, iron rails had stitched the US back together, albeit not across the north-south divide that needed mending the most.

Stanford's message to President Grant, like most media coverage of the event, portrayed the railroad as a triumph of American engineering and a harbinger of greater metallurgical wonders to come. The focus on iron and steel was understandable, given that railroads represented one of the main transformative technologies of their time. Amid the jubilant speeches and champagne toasts, however, nobody—from the White House to the speakers on stage—acknowledged that the material the Chinese laborers used the most was neither iron nor steel, gold nor silver. Few noticed that beneath their polished shoes lay thousands of humble wooden railroad crossties milled from evergreens whose sacrifice made such glory possible.

Railroads, symbolized by steel and steam, depended above all on a huge quantity of wood, surpassing the amount of metal used in train cars, bridges, telegraph poles, stations, and platforms combined. For most of the nineteenth century, the need for crossties made railroads the single largest consumer of wood in the country. To best support the rails and cushion trains roaring across the continent, crossties were historically made according to standard specifications: eight-foot-long rectangular wooden beams. Even today, 95 percent of railroad ties in America remain solid wood, resisting obsolescence in ways that wooden trestle bridges or railcars could not. Over the last two centuries, more than twenty-five hundred patents for newfangled crossties made from various composite materials have been issued, but so far none has matched the simple wooden tie's supreme combination of strength, spike-holding ability, and longevity. John Fristoe, the first president of the Railway Tie Association, famously declared, "A tree could plan no greater destiny than to give up its existence, not in old age but in its prime for the service of mankind."

In the early days of railroad construction, engineers sourced ties locally from whatever timber was close at hand. In the Great Basin area of the eastern Sierra Nevada, railroad crews targeted sturdy lodgepole pines (*Pinus contorta*), clearing entire hillsides for ties and timber bridges. Along their route, they erected temporary settlements, or "road ranches," creating strings of makeshift towns like knots tied loosely in rope, offering brief respite to weary railroad crews and their draft animals. Once track construction advanced toward the next terminus, these fleeting communities disappeared as swiftly as the forests that had sustained them, leaving behind ghost towns and tumbleweeds.

Similar scenes unfolded in forests nationwide as railroad tracks pushed farther into America's wilderness. As early as 1866, distinguished horticulturist Andrew Fuller was among the first to publicly call on railroad companies to address the devastating impacts of their expansion, writing: "Even where railroads have penetrated regions abundantly supplied, we soon find that all along its track timber soon becomes scarce. For every railroad in the country requires a continued forest from one end to the other of its lines to supply it with ties, fuel, and lumber for building their cars."

Equally destructive was the railroad's impact on Indigenous peoples who had called the western forests home for at least six thousand years before locomotives arrived. For the Washoe, Paiute, and Shoshone nations who considered these sacred landscapes central to their survival, the arrival of railroad crews disrupted the symbiotic relationship they had with the land. The Washoe people saw their forests stripped of ancient evergreens, effectively ending the annual pine nut harvests integral to community life. Paiute villages were demolished to make way for railroad tracks. Shoshone hunting grounds vanished overnight, replaced by sawmills and sprawling timber camps. The tribes' profound shock found little acknowledgment among railroad executives, whose gaze rarely extended beyond their own immediate economic interests. Displaced Indigenous communities were dismissed as barriers to progress, their dependence on the land ignored in the rush to bind the continent in rails.

But this early ransacking of lodgepole pines forced railroad barons to seek bigger and more reliable timber sources. Ambitious plans emerged to lay an additional 280,000 miles of track across the continent by the early twentieth century, and timber demand soon outpaced available forests alongside existing tracks. Beneath every mile of rail, twenty-five hundred wooden crossties lay perfectly spaced, meaning the planned new track construction would require about a quarter billion ties to complete. In the nineteenth century, most railroad ties were untreated with sealants to protect them from constant contact with moist soil, and thus they lasted only five to seven years. As a result, railroads needed to replace twenty million ties annually just to maintain existing tracks.

Not just any slab of timber, however, was suitable for a railroad tie. The wood needed to be capable of resisting decay, holding iron spikes, and withstanding the relentless pounding of passing trains without splintering. Railroad engineers felt that the highest-quality ties came from California's giant sequoias, whose durable heartwood, rich in natural tannins, repelled fungus and bacteria, rendering it remarkably impervious to rot. Shrink-proof, insect-proof, and practically everlasting, redwood earned extravagant praise from the California Redwood Association, which advertised it as "nature's lumber masterpiece."

Railroads were not alone in coveting redwood lumber. In bustling nineteenth-century New York City, builders installed redwood water towers on rooftops to regulate water pressure in newly constructed high-rises. The California Gold Rush, too, consumed vast quantities of redwood for flumes, sluice boxes, and mine-shaft supports to expand mining operations. Beyond these industries, redwood became California's go-to building stock, a cheap and convenient material that permeated nearly every aspect of daily life. Californians attended classes in redwood schoolhouses, sat at desks crafted from redwood, wore shoes whose leather had been tanned in redwood vats, and drank wine that was fermented in redwood barrels. Builders employed redwood extensively in homes, bridges, barns, and pipelines, as well as in sidewalks, gutters, and roads before concrete became commonplace. At the height of its popularity, some of the most tightly grained and richly colored wood ever milled on this planet was made into funeral caskets and buried with the dead six feet under—a final irony reflecting America's uneasy relationship with impermanence.

It became evident that California's sequoia groves promised unsurpassed wealth to anyone brave enough to cut them down. The timber rush sequoias created was reminiscent of the economic boom that had leveled the South's longleaf pine forests, except sequoias represented a vastly greater ecological jackpot—not necessarily because more trees would be cut down, but because they contained an unprecedented volume of biomass. The temperate rainforests along the Pacific coast held more living matter per acre than any ecosystem on Earth, surpassing the Amazon rainforest. A single giant sequoia could yield over five hundred thousand board feet of lumber, enough to build thirty-five houses, and seven and a half times more lumber than an entire acre of Lake States pines. The scale was almost incomprehensible.

As word spread, tens of thousands of lumberjacks left the declining pine camps of the Lake States and the South and rushed westward. California's Gold Rush had already lured thousands more miners eager to strike it rich, but gold proved elusive. Within months, most fortune-seekers abandoned mining and fanned out across the western territories in search of work. This sudden influx of unemployed men coincided with an emerging political strategy to populate the West, coupled with millions of acres of rocky, seemingly worthless land recently seized from Indigenous peoples. Booming railroad expansion, East Coast construction demands, the mining industry's hunger for timber, and the spectacular allure of recently "discovered" sequoias all converged simultaneously. Politicians saw an opportunity to capitalize.

In 1878, President Rutherford B. Hayes signed the Timber and Stone Act into law, offering 160-acre parcels of surveyed public lands that were deemed unfit for farming to any willing buyer at the astonishingly low price of $2.50 per acre (approximately $3,100 per acre today). With sequoias suddenly accessible and essentially free, thousands of speculators scrambled to claim the choicest groves. Land brokers facilitated rapid-fire transactions, and railroads snapped up vast tracts of public forest through dubious land grants and questionable interpretations of treaties with Native American tribes. The Timber and Stone Act thus invited industrial-scale logging on a level America had never witnessed.

Former gold miners and seasoned lumberjacks streamed into California's fog-shrouded coastal forests and the steep valleys of the Sierra Nevada. What these eager men found waiting for them was not the familiar landscape of Midwestern pines, but a forest so daunting that they hardly knew where to begin.

"The great size of the timber and the thick growth of underbrush have been sadly against us," wrote one overwhelmed pioneer confronting the ancient Pacific giants for the first time.

It was almost too much of a good thing.

12

Terrestrial Whaling

▲ ▲ ▲

LOGGING HAS ALWAYS ranked among the world's most dangerous professions. During the sequoia boom of the late nineteenth and early twentieth centuries, the fledgling US Bureau of Labor Statistics noted with alarm that no industry rivaled logging for its sheer lethality. A lumberjack working a forty-year career faced a fifty-fifty chance of dying beneath the trees. Hundreds of determined, low-paid men died each year cutting down trees whose scale defied comprehension, using primitive machinery and crude hand tools to wrestle behemoths that could, and regularly did, crush them like ants.

Day after punishing day, lumberjacks endured twelve-hour shifts, battling relentless rainstorms and fierce blizzards, risking hypothermia, frostbite, and dismemberment. Broken bones were routine, shattered kneecaps and noses commonplace. In these remote and unforgiving woods, injuries resembled wartime casualties. Logging camp doctors, often poorly trained, resorted to crude battlefield-style medicine, performing hurried amputations and stitching wounds with horsehair thread. They kept rough ledgers of injuries, their pages filled with grimly laconic entries such as "leg shattered," "eye punctured," or simply "crushed." A simple slip on a patch of wet moss could prove lethal. One unfortunate logger fell onto a broken branch and impaled himself so grotesquely it entered his rectum and exited his abdomen. When a lumberjack died in the field, his body would be set aside until the work shift was complete, before being carried out like a sack of flour at day's end. Sometimes fallen comrades pinned beneath logs too heavy to move were covered in branches and left behind out of grim respect. The sheer frequency and brutality of such injuries cast giant sequoias in a sobering light. As one logger with thirty-five years of experience noted, "It's not a job you want to bullshit your way into."

Observers often remarked upon the similarity between toppling sequoias and the dramatic spectacle of a whale hunt. Both required precise teamwork, entailed extreme danger, and ended with the death of a creature whose size defied biological limits. Lumberjacks referred to particularly formidable trees as "Leviathans," further reinforcing the maritime analogy. And when a sequoia weighing more than six humpback whales finally yielded and toppled, it emitted a sternum-shuddering groan so profound that loggers did not just hear it; they felt it in their bones. The act of bringing down a sequoia evoked visceral sensations akin to those often reported among whaling crews: primal, exhilarating, and terrifying in equal measure.

The methods used to fell these giants changed remarkably little over generations. The fundamental challenge was always the same: to turn a vertical stick into a horizontal one without getting killed. To accomplish this, loggers relied on brutally effective tools, including four-pound double-bladed axes that required frequent sharpening by dedicated smiths who kept grinding wheels spinning late into the night. Lumberjacks also carried fifteen-foot-long two-man crosscut saws ominously known as "misery whips," which they carefully coated in rendered bear fat to prevent rust and keep them gliding smoothly through the sequoias' fibrous wood. Loggers often gave their misery whips names or adorned them with painted designs, reflecting a peculiar reverence for the tools of their trade.

The trees complicated matters with their enormous buttresses, thick roots that flared outward like bell-bottomed jeans and often stood taller than a grown man. Below ground lurked another danger hidden within the intricate structure of sequoia roots. Remarkably shallow for trees so enormous, their roots penetrated only six to twelve feet beneath the soil but spread horizontally up to eighty feet, forming interlocking subterranean networks with neighboring trees. This rootbound stability caused some trees to remain defiantly upright after being mostly severed, forcing loggers to wait for a powerful windstorm to finish what their axes and saws began.

Before axes struck bark, specialized teams spent days building elevated platforms, or springboards, by chopping footholds into the trunks ten or twenty feet above the buttresses. From these precarious perches, they hacked deep, triangular notches into the enormous trunks, then drove steel wedges into the notches, forcing the trees to lean and teeter. The entire process could consume a full week, each blow resonating through the forest like the steady tolling of a bell.

Felling sequoias was merely the first, and shortest, phase in an arduous journey that began with rugged men hacking trails through roadless wilderness and ended when neatly hewn crossties arrived at railroad terminals half a continent away. Despite the best efforts of crews that cleared smaller firs and yews in the tree's projected fall path to create a cushioning bed, many of the behemoths shattered violently upon impact with the ground, their brittle wood exploding into thousands of useless shards. Those that survived intact posed an almost insurmountable challenge on mountainsides with perilous forty-five-degree slopes. A single sequoia trunk, slick with moss, weighed upwards of 250 tons, roughly five times heavier than old-growth pines back east. Loggers routinely abandoned perfectly usable logs whenever transportation became too difficult or costly, confident another giant stood just around the corner. Such casual disregard typified this era's prevailing attitude: an illusion of limitless natural abundance that justified staggering waste.

Once horizontal, sequoias required further specialized handling. Workers called ringers marked trunks at intervals, providing footholds for peelers, who used chisel-headed crowbars to strip away the bark that would otherwise gum up sawmill blades. Next came the buckers, who deployed misery whips to lop off limbs as thick as full-grown pines. When these formidable blades could not span a trunk's width, "powder monkeys" stepped in, expertly placing dynamite charges to blast the timber into manageable sections. If miscalculations caused explosions that destroyed adjacent trees, crews picked through mountains of debris to salvage what little usable timber they could for fence posts or shingles. Then they pressed onward, extending rail lines deeper into pristine sequoia groves and leaving mangled giants scattered in their wake.

Dragging logs out of these rugged landscapes demanded yet more ingenuity. Specialized workers called hook tenders secured logs with steel cables, connecting them into trains of timber known as log strings. A powerful machine called a steam donkey, invented by Eureka mill owner John Dolbeer in 1881, revolutionized this process. Operated by a skilled driver, this portable steam-powered winch dragged logs uphill and through treacherous terrain, assisted by the vigilant spool tender, who managed cables, and a donkey tender, who kept the boiler supplied with firewood and water. Logs were then pulled along crude, lubricated "skid roads" by teams of oxen managed by a highly paid "bullwhacker," assisted by a "waterslinger," who slicked the skid road with animal fat, allowing logs to slide forward more easily. Eventually, logs reached narrow-gauge railways—called skunks, for their smoky coal-powered engines—which carried them to flumes or railcars waiting below. The flumes themselves were engineering marvels: elevated wooden channels supported by intricate trestle systems extending miles over treacherous canyons and clinging precariously to cliff faces. Logs hurtled down flumes at alarming speeds, occasionally launching airborne on sharp turns and smashing through the structures, necessitating constant repairs.

Layers of technical expertise and complex logistics made the entire sequoia logging enterprise both astonishingly inventive and spectacularly wasteful. This panorama of contradictions was captured in the first report of the California State Board of Forestry, an advocacy commission created by the legislature in 1885. The board estimated that on average only 29 percent of any felled sequoia actually made it to market as finished lumber. The remaining 71 percent, including stumps, green tops, small branches, and stripped bark, went nowhere. And once any merchantable timber arrived at a sawmill, the circular saw blades cut kerfs three-eighths of an inch wide, turning 40 percent of the milled log into useless sawdust that was either burned or dumped into river valleys.

"Later, conveyors transported all the fine bits, the low-grade material, and the discarded short and odd lengths to a constantly burning bonfire or furnace," explained University of Pennsylvania historian Jared Farmer in *Trees in Paradise*. "The going price for California lumber," he added, "did not justify greater efficiency."

Walter Fry and John White captured the magnitude of these inefficiencies in the landmark book *Big Trees*: "There must be millions and millions of feet of lumber in the boles and limbs of sequoias lying on the ground . . . and there are millions more in the chutes and trestles upon which the logs were conveyed to the mills. . . . miles and miles of such trestles and chutes—miles and miles and miles. Waste, waste, waste."

One by one, ancient trees cascaded down mountainsides, leaving only thirty-foot-wide stumps as evidence of their existence. Sequoias once considered as exhaustless as the ocean were melting away, and each one toppled offered a contradiction. Its spiritual grandeur clashed with its brutal commodification, its timeless majesty severed by instantaneous destruction. Despite never setting foot in California's sequoia groves, Walt Whitman poignantly captured the somber ambivalence of logging them in his 1873 poem "Song of the Redwood-Tree."

> *Farewell, my brethren,*
> *Farewell, O earth and sky—farewell, ye neighboring waters;*
> *My time has ended, my term has come.*

13

Where Fools Rush In

▲ ▲ ▲

FACED WITH SUCH staggering waste and hazardous working conditions, it is hard not to wonder what drove men to enter the logging profession at all. Despite the risks, hundreds of thousands flocked to the sequoia groves between 1862 and 1900 with a fervor that defied logic, driven by impulses stronger than mere economic necessity or the promise of adventure. As logging towns sprang up, hidden worlds took root under the shadows of giant evergreens, worlds that contradicted the comfortable myths America preferred to tell itself. Tales of these secret societies occasionally surfaced in whispered rumors or sensationalized newspaper accounts, but not every story of the lumberjacks was one America wanted told.

Photographers extensively documented the punishing realities of sequoia logging, creating iconic images that captivated imaginations worldwide. Lumberjacks stood in heroic poses, dwarfed by colossal tree trunks, confidently gripping saws and axes, sporting thick handlebar mustaches beneath felt fedoras, their sturdy boots planted firmly, muscles taut under Stanfield's union suits and suspenders. From these romantic portrayals arose the legendary figure of Paul Bunyan, a larger-than-life folk hero born of working-class men spinning exaggerated stories around frontier campfires. These tales, originally oral folklore shared among loggers in Michigan and Wisconsin, spread westward and became nationally recognized expressions of American determination and frontier spirit. Bunyan's impossible hypermasculine feats—chief among them, leveling entire forests with a single swing of his axe—were the subject of seventeen popular books, which firmly established his place in American mythology. His legend grew especially prominent through promotional pamphlets distributed on behalf of a logging company in the early twentieth century, attempting to glorify lumberjacks' dangerous work and improve the timber industry's public image.

But if Paul Bunyan alone defined our full understanding of lumberjacks, they might be mistaken for proud, self-made men, boldly steering their lives and powering America's prosperity through a brotherhood forged in hazardous work. The truth, however, is more complex, more human, and far more unsettling.

The extractive industries driving America's westward expansion— logging, mining, agriculture, and fishing—employed transient laborers called bindlestiffs, who carried their few belongings in makeshift bedrolls (a.k.a. bindles) as they drifted from job to job along newly built railroads. These were invisible, disposable men, valued only as long as their bodies remained useful. Many had fled harsh economic conditions in urban centers or farms, often burdened by debt, seeking at least a temporary sense of stability in the forestry sector. Though the lumberjack's life offered little more than hardship and fleeting wages, for thousands of men it was still preferable to the poverty, the violence, or the crushing monotony they had left behind.

Rexford Tugwell, a Columbia University economist who later became governor of Puerto Rico and a top advisor to President Franklin Roosevelt, described Pacific coast logging communities as "a sad travesty at best." He painted the typical bindlestiff as "a rather pathetic figure, plodding along a woods-trail . . . Perhaps he drags along a frayed old imitation-leather suit-case; more likely everything in the world he can call his own is wrapped in a filthy blanket-roll that hangs upon his back. His eyes are dull and reddened; his joints are stiff with the rheumatism almost universal in the wettest climate in the world; his teeth are rotting; he is racked with strange diseases and tortured by unrealized dreams that haunt his soul."

Timber barons adhered strictly to prevailing gender norms and employed an exclusively male workforce, over three-quarters of whom were unmarried, reflecting their unstable and rootless existence. While most loggers were White American-born men, the industry also drew diverse groups—African Americans, Native Americans, Latin Americans, and Pacific Islanders—all bound together by poverty and desperation. They migrated westward sharing the same bleak reality of a profession that rewarded destruction rather than creation. With few employment options and scant dignity, these young men willingly traded safety for wages, convinced the forest's dangers paled in comparison to the threats they faced elsewhere.

The exploitation of sequoias also drew significant numbers of Chinese immigrants to logging camps along the Pacific coast. With their railroad-building skills, these men became indispensable to timber companies, constructing narrow-gauge rail tracks, felling trees, and working tirelessly in sawmills. Despite their critical contributions to California's economy, they faced severe discrimination and hostility. In 1885, Humboldt County systematically expelled hundreds of Chinese laborers, blaming them for depressed wages and brutal working conditions that were, in fact, enforced by the timber barons themselves. Local business directories proudly proclaimed Humboldt County as the only county in California "containing no Chinamen," illustrating the cruel irony of treating an essential workforce with spite.

Between logging seasons, lumberjacks flocked to San Francisco, squandering their paychecks on alcohol, gambling, and prostitution in a part of town aptly called Skid Row, named after skid roads. Newspapers of the era reported on frequent brawls, opium dens, and vice raids in the district, further cementing Skid Row's notorious reputation for moral decline. This neighborhood eventually became today's Tenderloin, a neighborhood still marked by some of the highest levels of homelessness, crime, and drug use in the city. On Skid Row, lumberjacks' rowdy excess could make sailors blush. But once the employment agencies had dispatched these men back to remote sequoia forests, they encountered something rarely experienced elsewhere: a semblance of home.

During the logging season, lumberjacks lived in rustic shantytowns with cookhouses at one end and oxen corrals at the other. The muddy ground between was littered with rusted wire, bent saw blades, discarded boots, torn shirts repurposed as rags, and broken rum barrels swarming with flies. Some camps built wanigans, floating bunkhouses set atop logs, which moved downstream alongside timber drives, providing even less stability but greater proximity to the shifting work front. Each camp typically housed around seventy lumberjacks. They slept two per bed without blankets on prickly hay mattresses inside clammy bunkhouses crudely constructed from rough-hewn planks. The dirt floors between beds were saturated with tobacco juice and blade-sharpening oil. Sopping leather boots piled around central woodstoves to dry, while sweat-stiffened clothing and tool belts hung from bunk ends. Without showers, toilets, or laundry facilities, the lumberjacks' collective masculine stench mingled with damp forest air. Accustomed to these conditions, they bore their discomfort with resigned stoicism, their hard lives etched into grimy faces and weary, bloodshot eyes.

Each morning, the cook's bugle jolted lumberjacks awake before dawn, sending them stumbling from bunks into bitter cold and darkness. They gnawed on cold salt pork, gulped down boiled coffee, and forced themselves to swallow multiple tall stacks of pancakes in enforced silence, as camp rules forbade conversation while eating. Not that it would have mattered, anyway. Their slang-filled English, shaped by prolonged isolation, was often incomprehensible.

Still in a half-asleep daze, with their axes tucked under their armpits to keep the metal from freezing, they trudged out into the forest gloom, standing shivering at the bases of sequoias until daylight pierced the canopy, signaling the start of endless chopping. They swung axes and pulled misery whips until twilight, then shuffled back to camp in exhaustion, wolfed down pork and watery bean slop, and collapsed into obliterated sleep, cradled in another man's arms for warmth. When rain seeped through planked walls and soaked their bedding, sleeping packed tight together to create frictional body heat provided their only hope of drying out by daybreak.

Camp life offered few comforts or distractions. These exclusively male environments rarely received visits from women, eliminating any heterosexual intimacy for nine months of the year. Pornography was strictly prohibited, along with any mail deemed "obscene, lewd or lascivious" under the 1873 Comstock Act. This federal law, championed by Anthony Comstock—a US postal inspector and secretary of the New York Society for the Suppression of Vice—dramatically expanded America's legal definition of obscenity. Comstock's aggressive moral crusade, thinly veiled in purported Christian ideology, led to thousands of arrests nationwide, even targeting something as innocent as a mailed love letter between consenting adults. Federal court records document at least eight criminal prosecutions involving "obscene love letters" sent to Pacific coast lumber camps. One notable case was *U.S. District Court vs. Lucile Dewar*, triggered by a letter Dewar mailed to Gerald Tomas, which began "Dearest Gerald—Hello Honey, will you make a date with me?" and concluded "Show this to all the boys down there." Comstock boasted of personally destroying millions of "obscene" images and books, illustrating the severe restrictions lumberjacks faced on basic entertainment or personal correspondence.

Yet one form of entertainment remained prevalent in the logging camps: same-sex intimacy. Alfred Kinsey's landmark 1948 study *Sexual Behavior in the Human Male* revealed that the highest incidence of all-male sexual activity occurred among "lumbermen, cattlemen, prospectors, miners, hunters, and others engaged in out-of-door occupations." The blandly written, eight-hundred-page scientific report immediately became a national sensation and sold over half a million copies. "Not since Gone With the Wind had booksellers seen anything like it," declared *Time* magazine. Kinsey's follow-up study, *Sexual Behavior in the Human Female*, garnered greater public attention, landing him on the cover of *Time* in August 1953 and placing the hidden sexual realities of American life squarely in the national spotlight. "In a society that deemed homosexuality either immoral, a sickness, or both," wrote Eric Cervini in his Pulitzer Prize–nominated *The Deviant's War*, "[Kinsey] provided evidence that gay sex was, in fact, objectively normal."

Kinsey viewed these same-sex encounters among lumbermen, cattlemen, and miners as situational rather than expressions of sexual identity, observing that the homosexual activity of hypermasculine lumberjacks "rarely conflicts with their heterosexual relations, and is quite without the argot, physical manifestations, and other affectations so often found in urban groups." According to Kinsey, sexual contact between men in logging camps arose strictly as a response to their prolonged separation from women. When asked why he engaged sexually with other men, one logger responded, "Cause there ain't enough women. If I can't get them, I've got to have the other." For many lumberjacks, these encounters reflected practical solutions for daily living rather than deeply embedded aspects of personal identity.

But if the purpose of history is to understand what truly occurred, the type of sexual intimacy Kinsey described has largely been erased, suppressed, and concealed. Because of societal taboos, explicit documentation of same-sex encounters among lumberjacks remains sparse, with primary sources mostly limited to oral histories, faded photographs, court records, and newspaper archives. Yet a growing body of scholarship has begun to illuminate previously unacknowledged histories of same-sex relationships and gender variance as integral parts of rural American life.

Washington State University historian Peter Boag documented multiple firsthand accounts of intimate relationships among migratory working-class men on the Pacific coast. One retired logger recalled that seven of nine men in his logging camp regularly had sex with each other. Another logger, describing his experience in a California gold mine where fifty-five miners lived and slept in close quarters, candidly admitted they often "grabbed it off by hand and got relief from one another."

An additional study of a Northern California logging camp from the late nineteenth century revealed how its isolation enabled lumberjacks to cast aside conventional gender norms. The high concentration of same-sex relationships fostered a normalizing effect similar to patterns documented among incarcerated men. Archival photographs captured lumberjacks carousing at events like the Miners' Ball, where half the men danced female roles, clad only in strategically placed patches signifying their assumed femininity. At spirited Cowboy Stag Dances, rugged lumberjacks swung each other through fast-paced square-dance do-si-dos. Many of these gatherings explicitly celebrated unions known as bachelor marriages, informally officiated by respected figures within the logging crews. These unions served both emotional and practical purposes as they secured companionship and mutual protection in harsh outdoor environments. Lumberjacks always worked in pairs to protect each other, with cutting partners during the day typically becoming bedroll companions at night.

Partnerships usually matched an older lumberjack in his thirties or forties, called a "wolf," with a younger logger, known as a "punk" or a "jocker" (a slang term referencing the penis). Wolves typically occupied more dominant, traditionally masculine roles and exerted a coercive or intimidating presence, clearly preferring sexual relationships with younger men. Jockers and punks, meanwhile, generally assumed receptive sexual roles and took responsibility for domestic tasks around camp, including building fires, sewing torn clothing, cleaning cookhouses, and even bathing and shaving their wolf partners. Some held specialized roles as "whistle punks," who blew whistles signaling meal breaks or impending tree falls. Over time, as younger loggers gained experience and confidence in their skills, they would eventually assume the role of wolf themselves, training and partnering with newer camp arrivals in a cycle that ensured continuity within the workforce.

In some cases, these Queer lumberjack relationships revealed the humorous resilience of working-class men responding to outsiders who misunderstood logging camp culture. One federal law enforcement agent investigating rumors of same-sex relations at a Pacific coast work camp was surprised by their playful reaction. In his official report, titled "Headquarters File no. 071.1," he wrote:

> To Chief Special Agent Olin Colip Anchorage.
> From W. M. Bettis Special Agent Fairbanks.
> Subject: Sodomy Healy Section House. . . .
>
> . . . I recieved information of an act of Sodomy in the Healy Section house [and] made a trip to Healy to investigate this roomer [sic]. . . . When I asked [the man] what happened to him he said . . . I was very drunk as I got to the bottom of the stairs I fell the next think I knew someone was pulling down my pants.
> This man . . . caught me from behind and performed an undecent [sic] act on me. I did not no what he was doing but I had a good idea. . . . I then asked [someone else] what he knew about this deal he said yes he saw it. I asked him why he did not stop it? He said well I sure got a hell of a kick out watching it I sure did enjoy it he also said it was the most fun I have had in three years.
> I then asked . . . if [he] wanted to place charges . . . [He] then grinned at me and said yes I am going to charge him $5.00 when I see him. . . .
> {signed} W. M. Bettis

Far from being a sporadic or rare social dynamic of logging camps, same-sex intimacy was simply embedded into everyday life. Yet these encounters did not align with contemporary notions of gay identity, largely because such identities did not yet exist. Even the term "heterosexuality" did not make its debut in Merriam-Webster's *New*

International Dictionary until 1923, and medical authorities of the era still debated whether homosexuality was a vice or mental illness, not considering it just another dimension of human diversity. Regardless of how these men perceived themselves, the comfort and companionship found through sexual intimacy or bachelor marriages was inevitably overshadowed by the pervasive fear of getting caught.

At the height of the sequoia logging era, police conducted frequent undercover stings across America, arresting over fifty thousand people each year—one arrest approximately every ten minutes—for suspected homosexual behaviors. Actions as innocuous as dancing or holding hands could lead to persecution, imprisonment, or worse. Suspected homosexuals were strictly barred from public-sector employment under federal and state anti-lewdness laws, wrongfully accused of causing "apprehension and revulsion" among their coworkers.

In the late nineteenth and early twentieth centuries, some of the harshest anti-homosexuality laws existed in the American West. California, Oregon, and Washington joined twenty-nine other states in authorizing forced castration, lobotomy, or lifelong confinement in mental institutions for anyone accused of homosexual acts. In Oregon alone, between 1917 and 1921, three hundred men were forcibly sterilized or institutionalized under state eugenics laws specifically targeting homosexuality. Roommates, landlords, and neighbors constantly watched for signs of homosexual activity, and anyone accused could expect to find their names printed in newspapers, publicly shaming them for their supposed deviance.

Daily threats of violence forced countless men to retreat further into society's margins, especially toward America's vast wilderness areas. Forests have long served as sanctuaries for Queer men, offering refuge from watchful eyes and judgment. Indeed, the word "forest" itself, derived from the Latin *foris*, meaning "outside," originally described "extensive tree-covered districts set aside for hunting" in medieval England. And what is cruising for sex if not a form of hunting?

Throughout history, forests around the world repeatedly became safe havens where men could temporarily shed the fear that drove them into hiding and embrace brief moments of intimacy beneath protective evergreen canopies. For those who might doubt this hidden facet of gay life, consider taking a closer look at the Scotch pine grove overlooking the men's bathing pond in London's Hampstead Heath, or Berlin's sprawling Tiergarten after dusk, where hundreds of men gather beneath monumental giant sequoias and dawn redwoods (*Metasequoia glyptostroboides*). Wander the shadowy pathways of the Meat Rack, a sunken pitch pine forest at Fire Island Pines, a haven for New York's gay community since 1920. Take a brisk walk through the marshes and dunes of Cape Cod National Seashore in Massachusetts, where ancient pitch pine scrublands conceal Provincetown's promiscuous Boy Beach. Stroll along the secluded trails of San Francisco's Buena Vista Park, cloaked in dense Hollywood junipers (*Juniperus chinensis*) and Torrey pines (*Pinus torreyana*), America's first, and still largest, gay cruising mecca. Hike through the riparian bottomlands above the Columbia River in Rooster Rock State Park near Portland, Oregon, or recall the forested wilderness once hidden within Atlanta's Piedmont Park, whose gay cruising grounds were bulldozed for urban development, extinguishing decades of private intimacy. Or consider joining the long line of men from every political persuasion who, since 1885, have met for sexual encounters after dark in the densely wooded Lafayette Square park, immediately adjacent to the White House in Washington, DC.

Even in New York City, a metropolis of concrete and skyscrapers, cruising flourishes in The Ramble, a thirty-six-acre woodland in Central Park designed by Frederick Law Olmsted to evoke an upstate forest. Its winding trails, dense vegetation, rustic bridges, and rock outcroppings make it famously labyrinthine. Since at least the early twentieth century, gay men have gathered there discreetly, drawn by its seclusion and natural beauty, yet always mindful of potential violence and arrest.

Some may find these revelations disconcerting. America's forests are more often remembered as the exclusive domain of mythical heroes, icons of strength, solitude, and untamed masculinity. Imagining Paul Bunyan acolytes engaging intimately in forests during America's pioneering days can feel akin to learning the Marlboro Man died from lung cancer.

As Clifford Westermeier, a former history professor at the University of Colorado Boulder, cautioned, "To tamper with the image of a folk hero, a historic formula, a legend, and most of all, that of the American cowboy heritage is probably more dangerous than the proverbial where 'fools rush in.'"

The proverb "fools rush in where angels fear to tread," from Alexander Pope's 1711 poem "An Essay on Criticism," warns against acting impulsively in dangerous situations that wiser souls instinctively avoid. The reckless spirit of the sequoia logging culture embodied this proverb precisely. Inexperienced or desperate men routinely confronted dangers that more prudent individuals—the angels—would have wisely shunned. Yet their quest for companionship repeatedly drew them deeper into forests, motivated by dreams of material wealth and belonging that rarely, if ever, came true.

In the end, despite unrestricted access to seemingly inexhaustible forests, sequoia logging never yielded sustained wealth. The industry demanded massive upfront investments in sawmills, railroads, and flumes, creating steep financial hurdles that timber companies struggled to overcome. The Timber and Stone Act had opened the land to too many small timber prospectors and sawmills, which inadvertently flooded regional markets with excess lumber. Chronic overproduction depressed prices, leading to bankruptcies piling up by the dozen. Like many extractive industries, sequoia logging lurched between dramatic booms and devastating busts, leaving behind ruined dreams rather than lasting fortunes.

By the time the last sawmill fell silent, perhaps 1.5 million ancient sequoias had been cut down, leaving fewer than 5 percent of their original population still standing. The ecological consequences were catastrophic: entire watersheds degraded, native species displaced, and vast storehouses of carbon lost. When interviewed decades later, former sequoia lumberjacks expressed genuine shock that the trees had vanished. Their surprise stemmed from a delusional optimism. Having witnessed firsthand how fast sequoias grew, they assumed new generations of trees would quickly replace the giants they felled, but failing to grasp that regenerating ancient forests required far longer timescales.

Ironically, the wealthy Victorians who visited William Hanford's London exhibition in 1853 succeeded where Americans had failed. Captivated by the enormous trees on display, Britons propagated over half a million giant sequoia and coast redwood seedlings across England, vastly outnumbering the mere eighty thousand sequoias that survived California's logging frenzy. Thus England flourished as a new Eden for sequoias, while America—ambitious, industrious, shortsighted—stood diminished, having squandered one of its most majestic natural inheritances. This imbalance highlighted another bitter irony: A single mature sequoia could produce up to four hundred thousand seeds annually, suggesting an immense ecological potential that had been recklessly discarded.

Even amid this ecological plundering, a group of five hundred former loggers sought redemption. In 1884, they established a utopian socialist community called the Kaweah Colony, the name meaning "here we rest." The colony attracted idealists who believed cooperative ownership could undo the wastefulness they had witnessed. Their socialist newspaper, *The Kaweah Commonwealth*, passionately advocated equal rights, worker protection, and conservation decades before these ideas entered mainstream discourse. For six brief years, these retired lumberjacks lived in humble squatters' cabins beneath the protective shadows of giant sequoias. Though their communal experiment ended with eviction when Sequoia National Park was established, their quiet endeavor affirmed that profits from evergreens could manifest as many things—including love.

14

Cut Out and Get Out

▲ ▲ ▲

WHILE CALIFORNIA'S ANCIENT sequoias fell with tragic finality, leaving valleys strewn with splintered stumps and shattered lives, a parallel drama was unfolding thousands of miles to the east. In the half-forgotten Lake States, timber barons raced to squeeze every last dollar out of the North Woods before these trees, too, faded into memory.

Just eighteen days after General Robert E. Lee surrendered to the Union at Appomattox Court House in 1865, Ezra Cornell made a fateful investment. Already wealthy from his partnership with Samuel Morse in the telegraph industry, Cornell turned his keen entrepreneurial attention toward timber. Exploiting an overlooked provision of President Abraham Lincoln's Land-Grant College Act of 1862, originally intended to fund colleges focused on agriculture and the mechanical arts, Cornell acquired five hundred thousand acres of prime northern Wisconsin forests from the federal government at the astonishingly low price of fifty cents per acre.

But this was no empty wilderness passively awaiting settlers' axes. These lands had been seized from the Menominee Nation through sixty-three coercive federal treaties, systematically reducing their homelands by 97 percent, from ten million acres to just 235,000 acres. The Menominee watched in grief and anger as outsiders began carving up their ancestral forests in a cycle of extraction and displacement that would reverberate for generations.

Cornell pledged all proceeds from his new Wisconsin timber holdings to endow his namesake university in Ithaca, New York. Having closely observed America's westward expansion, Cornell shrewdly anticipated that the Lake States' forests would skyrocket in value, urging university trustees to resist any impulse toward immediate liquidation. Unlike most land-grant institutions that promptly sold their federally granted parcels for quick profit, Cornell University appointed agents in Wisconsin to oversee logging operations directly. These emissaries arranged timber auctions, negotiated contracts, and arbitrated disputes among rival logging firms.

To the university trustees, this cautious pace represented prudent financial stewardship. Critics, however, condemned the university as a distant landlord, indifferent to local environmental destruction and community upheaval. Despite the controversy, Cornell's instincts proved impeccable. Postwar Reconstruction spurred explosive economic growth, turning the Lake States timberlands into some of the most valuable real estate in the world.

Cornell himself did not live to see his investment bear fruit, but in 1875 the university reaped a remarkable windfall. It sold a fifty-thousand-acre parcel to Frederick Weyerhaeuser, an ambitious forty-one-year-old German immigrant who had risen from penniless brewery worker to Midwest lumber magnate. Already respected for his sharp eye for timber opportunities, Weyerhaeuser recognized the tremendous wealth hidden in Wisconsin's forests and was willing to pay handsomely for it. Cornell University's return on the deal marked one of the most successful episodes of land speculation in US history: a twentyfold increase, yielding an endowment of $5 million (about $146 million today).

The deal's financing terms compelled Weyerhaeuser and his investors to repay the enormous sum within ten years at 7 percent interest—conditions that forced them to accelerate logging to meet their obligations. This aggressive strategy of forest liquidation, aptly called "cut out and get out," generated immense fortunes for both Cornell and Weyerhaeuser. Federal tax incentives at the time further encouraged timber companies to liquidate woodlands, unconstrained by environmental concerns.

Weyerhaeuser's lucrative strategy attracted immediate imitators, most notably Henry Ford, who purchased 313,000 acres in Michigan's Upper Peninsula from the federal government—land ceded by the Ojibwe and Chippewa nations. Ford clear-cut the region to mass-produce wooden body frames for his revolutionary 1908 Model T. This timber acquisition helped lower the Model T's price in future years below $300, permanently reshaping American transportation and urban life.

Across Wisconsin and Michigan's Lower Peninsula, sawmill owners scrambled to reach previously inaccessible white pines, hastily laying new rail lines that penetrated thousands of acres of remote northern forests. Railroads soon connected Milwaukee and Chicago, principal ports shipping pine eastward via the Erie Canal. Yet the timber boom eventually descended into logistical chaos. Independent contractors, sawmill operators, and railroad managers relied on tenuous handshake agreements and erratic intermediaries, causing bottlenecks and confusion. Timber frequently disappeared or unexpectedly changed hands due to theft or miscommunication, sparking violent confrontations between rival crews. Competing railroads adopted incompatible track gauges, forcing lumbermen into costly, time-consuming freight transfers. Valuable logs often sat idle for months, tying up capital and strangling profits.

Resolving this turmoil required timber barons to adopt a radical new business model: vertical integration. Rather than relying on loosely connected operations, owners like Weyerhaeuser and Ford consolidated every step of the supply chain under their direct control—from standing forests to sawmills, railroad lines, and finished lumber distribution. With this shift, the full force of industrial capitalism arrived in America's North Woods.

Driven by soaring immigration, urbanization in the Midwest, and the post–Civil War Reconstruction building boom in the South, lumber production reached staggering proportions. Legions of ravenous loggers descended upon the North Woods, wielding transformational technologies like steam engines, circular saws, and railroads to wrestle tens of millions of trees to the ground. Pines were sawn into two-by-fours, split into shingles, and hammered into balloon-framed houses nationwide. America built itself in the grand style of Europe's stone cities, but almost entirely from wood, elevating timber to the nation's largest industry, surpassing cotton, textiles, and steel in economic impact. Lake States pine harvests peaked in 1890 at ten billion board feet annually—enough wood to build an inch-thick, fourteen-foot-wide boardwalk circling the earth's equator five times.

James Little, a prominent lumberman and conservationist, described the furious pace of logging as "not only burning the candle at both ends . . . but cutting it in two, and setting the match to the four ends to enable them to double the process of exhaustion."

The first historian of lumbering in the North Woods, George Hotchkiss, calculated in 1898 that the Lake States lumber industry produced $4 billion in raw materials to build the nation during the half century after Appomattox—an amount almost double what the Union spent fighting the Civil War and three times the value of all the gold mined in California during the same period. Yet these profits did not benefit everyone equally, if at all. By the early twentieth century, one-third of Menominee's tribal members lived below the poverty line, while Cornell University's endowment swelled into the billions and Weyerhaeuser's market capitalization surpassed that of John D. Rockefeller's Standard Oil Company.

Like drinking water from a fire hose, a young nation in the throes of exponential growth had swallowed up nearly a million square miles of northern evergreens, leaving behind an uninhabitable, pockmarked "stump prairie" wasteland. Reflecting soberly years later, one of Weyerhaeuser's former business partners confessed, "We wiped [the forest] out in an eyewink of history and left behind fifty years of heartbreak and economic ruin."

15

Tinderbox

▲ ▲ ▲

WHILE LAKE STATES timber barons converted evergreens into fortunes, they left behind a deadly legacy. Loggers habitually discarded smaller trees, branches, and broken limbs into jagged heaps known as slash piles, forming chaotic tangles of dried needles, resin-heavy bark, and wood splinters. Sunbaked over months into brittle tinderboxes, slash piles could rise ten feet high and stretch uninterrupted for acres, creating fuel loads far more volatile than natural forest understory. As railroads extended their winding tentacles deeper into America's heartland, these slash piles proliferated ominously along the tracks, ready to ignite at the faintest spark from passing trains.

More alarming than steam locomotives' prodigious wood consumption was their propensity to catch *other* things on fire—from the passengers themselves to the train cars they pulled and the countryside they rolled through. Black smoke and hot cinders billowed from smokestacks along entire train lengths, showering helpless passengers who cowered in their seats. On a star-studded American reading tour, Charles Dickens recalled being enveloped in "a whirlwind of bright sparks, which showered about us like a storm of fiery snow," underscoring the inescapable perils of rail travel.

These conditions culminated disastrously on October 8, 1871, when the deadliest wildfire in American history engulfed Peshtigo, Wisconsin, and burned over one million acres in a matter of hours. The Peshtigo firestorm moved so violently that tornadoes of flame spun through the air, igniting buildings hundreds of yards ahead.

Survivors recounted how the heat grew so intensely it melted sand into glass and fused metal tools into twisted sculptures. The fire claimed an estimated twenty-five hundred lives, though precise numbers remain uncertain since many of the victims were buried anonymously in mass graves.

In a cruel twist of fate, on the same night another blaze roared to life just 250 miles to the south: the Great Chicago Fire, which killed three hundred people and destroyed seventeen thousand buildings. As the fire surged through the city, it gained strength from the streets themselves, which were paved with hydrocarbon-rich pine blocks soaked in highly flammable coal tar. Two cities nearly erased in a single day offered a wake-up call to a nation now forced to confront the perilous consequences of its ambition.

Yet history repeated itself with grim inevitability when sparks from railroad tracks ignited slash piles amid dense stands of eastern white pine, unleashing the 1894 fire in Hinckley, Minnesota. Flames obliterated a dozen towns and claimed over four hundred lives, still ranking as the third-deadliest wildfire in US history.

In a bold response to rampant wildfires, President Theodore Roosevelt established the US Forest Service in 1905, marking the first comprehensive federal effort to manage and protect America's national forests. Roosevelt appointed as its first director Gifford Pinchot, founder of Yale University's forestry program and an influential advocate for sustainable forest management. Under Pinchot's guidance, Roosevelt expanded the nation's protected lands from forty-three million acres to nearly two hundred million, effectively launching the modern American conservation movement.

Pinchot's legacy was undoubtedly complicated, however. Alongside his celebrated forestry work, he became deeply entwined in the troubling ideology of early twentieth-century eugenics. He compared human "races" to tree species and openly promoted selective breeding, frequently espousing disturbing claims that just as forests required careful management, so too did human populations. Among his less controversial views, Pinchot clearly understood the existential threat posed by wildfire, writing, "Of all the foes which attack the woodlands of North America, no other is so terrible as fire."

Long before devastating infernos engulfed Midwestern forests, Indigenous peoples across North America—from the Algonquian tribes on the Atlantic coast to the Menominee in Wisconsin and the Miwok in California's Sierra Nevada—had fostered a sophisticated relationship with fire. Through generations of observation, Native Americans had identified the conditions that bred wildfires: dry grasses, fallen branches, dense undergrowth, and strong winds. By strategically setting controlled burns during cooler, wetter periods in autumn, they cleared excess vegetation and reduced fuel loads. The smoke from these low-intensity fires produced a characteristic autumn haze, spawning the now-familiar expression "Indian summer."

Peter Del Tredici, a scientist formerly at Harvard University's Arnold Arboretum, correlated fire scars on older trees with nearby younger growth, concluding that Native Americans typically performed regenerative burns about once every twenty-five years. Elders collectively decided which woodlands needed burning, demonstrating a profound communal knowledge and deliberate forest management strategy.

This ancient practice, today called "cultural burning," "prescribed burning," or "light burning," created dynamic mosaics of forest in varying stages of regrowth, providing abundant resources for food, tools, and shelter. New plant shoots stimulated by light burning attracted game animals such as deer and elk, ensuring sustainable hunting opportunities. By embracing fire with intention, Indigenous peoples safeguarded their forests, communities, and livelihoods from the destructive effects of wildfires.

Initially, when the Forest Service was established, there was some openness toward Indigenous fire-management practices, already employed by a handful of White farmers and ranchers. But attitudes shifted in August 1910, when the Big Burn swept across the Northwest, just a few months after President William Howard Taft fired Gifford Pinchot over a policy dispute. This conflagration consumed three million acres across Idaho, Montana, and Washington, making it one of the largest wildfires by land area ever documented on US soil. Local Idaho newspapers attributed the inferno to dry slash piles alongside railroad tracks and the "flammable sap" of Idaho's western white pines (*Pinus monticola*).

Idaho author Betty Goodwin Spencer described forests that "staggered, rocked, exploded, and then shriveled under the holocaust . . . [with] great, red balls of fire [that] rolled up the mountainsides." Hurricane-force winds exceeding seventy miles per hour turned scattered regional fires into monstrous cyclones of flame so intense that evergreens exploded like artillery shells. Firefighters reported terrifying sounds akin to locomotives roaring overhead as swirling vortices uprooted mature, hundred-foot-tall western white pines and tossed them through the air like matchsticks. The Big Burn incinerated enough timber to fill a freight train twenty-four hundred miles long and reduced one-third of Wallace, Idaho, to ash. Smoke darkened skies all the way to New York. Ultimately, seventy-eight firefighters perished in America's deadliest wildland firefighting incident.

The fallen firefighters became national heroes, their sacrifice enshrined in popular memory and official policy. Public outrage surged. Politicians demanded answers. Forest Service officials defended their agency vigorously, arguing that inadequate staffing and a lack of funding had limited their ability to address a fire of such scope. In response, Congress doubled the agency's budget and dramatically expanded its firefighting force. This newfound strength came with a hardline ideology of total fire suppression, where even minor, ecologically beneficial fires were extinguished immediately, and traditional Indigenous practices of light burning were banned.

Despite pleas from forestry experts and ranchers to reconsider Indigenous wisdom about controlled burns, Forest Service chief Henry Graves was unequivocal in his rejection. "The doctrine of light burning . . . is nothing less than the advocacy of forest destruction," Graves said, "and those who preach the doctrine have a large share of responsibility for fires which their influence has caused." His leadership cemented a rigid institutional stance that would persist for decades.

Thus began America's war against wildfire: a crusade rooted in the conviction that every fire was inherently harmful to the national interest. Fire had become a formidable adversary that America was determined to eradicate, no matter the price. But this campaign went beyond ecological mismanagement; it became a tool of cultural erasure, criminalizing practices that had safeguarded forests for generations.

By 1924, when Native Americans were belatedly granted US citizenship, the federal government had already seized nearly two-thirds of their ancestral lands through a series of three hundred coercive treaties justified by the 1887 Dawes Act. Signed by President Grover Cleveland, this law dismantled tribal control of reservation lands by reallocating communal territory into individual parcels. Under this policy, Indigenous heads of households received 160 acres each, single men and women received 80 acres, and children under twenty-one received 40 acres. Conveniently for Anglo-American interests, any reservation lands that remained unclaimed after processing the allocations were opened for sale to White settlers. When Native leaders refused to relinquish territories, the US military enforced compliance through threats of violence.

Religious advocates framed the Dawes Act as a humanitarian measure intended to bridge what they viewed as a vast cultural divide between "barbarism and civilization." In reality, the law operated alongside other oppressive federal policies designed to eradicate Indigenous cultural traditions—particularly those practices essential for maintaining forest health through controlled burns. Foremost among these was the 1902 "haircut order," issued by the Bureau of Indian Affairs, which withheld food rations from Native men who refused "the wearing of short hair," adding that short hair would "certainly hasten their progress toward civilization."

Although some reform-minded Americans genuinely believed individual land ownership would inspire Indigenous communities toward productivity and Christian virtues, many Indigenous peoples fiercely resisted these policies as yet another form of colonial domination rather than the benevolent assimilation proclaimed by the policies' architects.

Ultimately, the Dawes Act proved disastrous. Across the West, Native Americans became confined to shrinking, bounded reservations, forced to watch helplessly as White settlers leveled their ancestral forests into towns and farmland. By 1912, newly established states like Arizona and New Mexico exemplified how efficiently this landscape transformation could occur, especially in regions most vulnerable to wildfire. Trapped between White expansion and their own disappearing cultures, Native American communities were compelled to abandon traditional land practices and enter the wage economy, further accelerating cultural dispossession.

The federal government's assault on Indigenous peoples' culture went further, striking at the very heart of their families. By 1925, the government had forcibly separated nearly half of all Native American children from their parents and placed them at nearly 150 government-owned Christian boarding schools around the country. Among these, the Carlisle Indian Industrial School in Pennsylvania stood as the prototype. Founded in 1879 by Captain Richard Henry Pratt, Carlisle operated under the chilling motto "Kill the Indian in him, and save the man."

At Carlisle, children from more than 140 tribes were stripped of their identities, forced to cut their hair, abandon their language, wear a military-style uniform, and adopt an Anglo-American name. Speaking their native tongue resulted in punishments like having their mouth washed out with soap or being subjected to solitary confinement.

These institutions, staffed and sanctioned by leaders of the Christian faith, deliberately aimed to shatter Indigenous identities—a mission they executed with vile proficiency. Records indicate that at least 186 children died at Carlisle due to disease, malnutrition, or abuse, and many were buried in unmarked graves, far from their homelands. The trauma inflicted by such schools has had lasting effects, contributing to intergenerational cycles of grief, loss, and cultural disconnection within Native communities.

These deliberate acts of cultural erasure complemented an emerging American vision of pristine wilderness, epitomized by the establishment of the National Park Service in 1916. Influenced by John Muir, revered as a founding father of American wilderness preservation, national parks became symbols of a romanticized notion: pure landscapes devoid of human presence. Muir championed a wilderness ideal shaped explicitly by White heterosexual notions of masculinity, celebrating "brave and manly and clean" mountaineers who, in his view, stood superior to urban dwellers who were "mildewed and dwarfed in disease and crime." His disdain for Indigenous peoples was explicit. In his 1911 memoir, *My First Summer in the Sierra*, Muir described Native Americans as "dark-eyed, dark-haired, half-happy savages," whose presence degraded otherwise unspoiled landscapes.

Muir's vision of an unpeopled wilderness justified the forced removal of Native Americans from ancestral lands and deeply influenced subsequent generations of naturalists, including Aldo Leopold, the celebrated forester-philosopher and author of *A Sand County Almanac*. Writing during the height of Jim Crow, in 1925, Leopold advocated for wilderness to be "segregated and preserved," even as federal policies continued to forcibly assimilate Native Americans into White society. The influential *Leopold Report* reinforced this racially charged ideology, describing national parks as "a vignette of primitive America," landscapes deliberately erased of human presence—further rationalizing the displacement of Indigenous communities whose stewardship had shaped these ecosystems long before European colonization.

But the national crusade against wildfire, rooted in the same logic that justified Indigenous erasure, did more than rewrite America's ecological story. It ushered in another profound shift in the foundations of American society at the dawn of the twentieth century. The wood-fueled locomotives, steamships, and industrial furnaces that had propelled the nation's westward ambition began giving way to a powerful new energy source: fossil fuels.

Coal, formed from Devonian-era Christmas trees fossilized in dense seams underground, offered a fuel far more potent and reliable than freshly split firewood. From 1870 to 1910, annual coal production surged from roughly forty million to nearly half a billion tons, transforming city skylines into dark silhouettes framed by smoke. The sharp pivot to coal was not merely a change in energy sources but represented a broader reorientation of the American economy away from its millennia-old reliance on forests toward a new dependence upon subterranean resources.

Wood, which had fueled domestic fireplaces, industrial machinery, and railroads, became increasingly obsolete in an age of coal-fired electricity. Oscar Wilde, ever attuned to irony, observed wryly in his 1890 novel, *The Picture of Dorian Gray*, that owning a coal mine afforded a gentleman "the decency of burning wood on his own hearth," highlighting how coal's dominance had turned traditional firewood from necessity into luxury.

Yet as America sought to distance itself from its ancient reliance on evergreens, it would soon discover that its uneasy relationship with fire was far from extinguished. Flames would return, summoned with a vengeance in a new age defined by fossil fuels, industrial might, and the relentless engines of modern warfare.

16

Airplane Spruce

▲ ▲ ▲

FIRE, HARNESSED FIRST from ancient evergreens and later from the coal seams beneath their fossilized remains, built civilizations, powered factories, and drove trains across continents. But fire also forged weapons—tanks, submarines, chemical explosives—that shattered nations. With the outbreak of World War I, humanity plunged into an era where coal, steel, and timber were no longer just fuels of domestic prosperity, but essential tools of mass destruction. Above the trenches of Europe, airplanes carved a terrifying new battlefield from the sky, forever changing the calculus of geopolitical conflict.

When the war erupted in 1914, airplanes were fragile novelties. Allied and Axis powers each possessed fewer than three hundred aircraft, all poorly armed, flimsy reconnaissance craft constructed largely from wood frames covered in linen fabric. But as aerial surveillance became critical, these delicate planes rapidly evolved from passive observers into deadly tactical instruments. Engineers raced to develop forward-firing machine guns synchronized to shoot between spinning propeller blades without striking them. This ingenious advancement transformed isolated dogfights into large-scale squadron battles, the rhythmic chatter of synchronized machine guns echoing above Europe's trenches.

Technological advances in Europe's aviation sector caught the United States flat-footed. Despite its vast industrial capacities, America had virtually no infrastructure or expertise in building sophisticated military aircraft. The few American-made warplanes had proved disastrously prone to spontaneous combustion, earning them the grim nickname among US Air Force pilots as "flaming coffins." Allied commanders thus discouraged American airplane manufacturing, urging it instead to focus on supplying crucial materials Europe lacked, especially timber. Resource extraction happened to be America's specialty.

Early aviation pioneers discovered that wood's remarkable balance of lightness, strength, and flexibility made it uniquely suited for aviation. Orville and Wilbur Wright spent four painstaking years developing their famous Wright Flyer, carefully testing thousands of wood samples. By the time they took flight in 1903, they had determined that spruce was the optimal choice for constructing wing spars and fuselage frames. In 1909, the US Forest Service established the Forest Products Laboratory at the University of Wisconsin–Madison, the first research center dedicated exclusively to testing wood's properties and commercial uses. Using a million-pound hydraulic press, scientists empirically confirmed what the Wright Brothers had intuitively known. Spruce possessed an exceptional strength-to-weight ratio (twenty-three pounds per cubic foot on average across five major types of spruce), which was almost three times stronger than steel at comparable weight, and significantly lighter than most conifers and hardwoods. Laboratory tests also revealed that spruce possessed unusually tough fibers, which resisted splintering even under gunfire—leading to the production of durable airplane propellers. When this data reached Allied military commanders, a French aviation officer remarked, "Your spruce will prove the decisive factor in the big European conflict."

As the world's fate hung in the balance, military strategists gathered in Washington, DC, to decide which spruce species to tap for the war effort. Engelmann spruce (*Picea engelmannii*) was dismissed due to numerous knots and its stunted growth in harsh Rocky Mountain winds. Norway spruce (*Picea abies*) proved notoriously difficult to mill smoothly and tended to drip resin unless thoroughly kiln-dried, rendering it unsuitable for mass production. The red spruce prevalent in Maine and New York (*Picea rubens*) rarely grew tall enough to meet aeronautical specifications, and mature red spruce forests had mostly been claimed by paper mills supplying media companies like *The New York Times.*

America's strategic choice narrowed to Sitka spruce (*Picea sitchensis*), an obscure evergreen conifer native to the coastal fog belt stretching from southern Oregon to Alaska's Aleutian Islands—a boulder-strewn landscape known as "west of west." Thriving primarily in low-lying saltwater bogs less than a mile from the ocean, Sitka spruce are impervious to fierce Pacific storms that ravage lesser species. Their needles evolved to capture moisture directly from the air, a crucial adaptation in this perpetually damp and mist-shrouded environment, where annual rainfall reaches eighteen feet. From a minuscule seed weighing just one-thirteen-thousandth of an ounce—a mere speck of dust—a Sitka spruce, given ideal conditions, can mature into a forest giant, towering up to three hundred feet and living more than eight centuries.

Despite their majestic stature, these ancient giants had historically been ignored by the timber industry. Home builders avoided spruce due to its reputation for decay, and the price spruce timber fetched rarely recouped the cost of hauling it out of the forest. More durable fir and cedar dominated lumber markets, leaving Sitka spruce essentially worthless, harvested only by necessity if it impeded access to more valuable species. Decades of selective logging had left prime spruce stands untouched and unreachable, lacking roads or rail infrastructure to support logging them. With World War I, however, "airplane spruce" suddenly rose from obscurity to preeminence, and America held a monopoly on the world's supply.

In spring of 1917, military delegations from Britain, France, and Italy visited the Pacific Northwest on an urgent mission to secure the cooperation of sawmills in turning out vast quantities of spruce for the war mobilization. Their request: 117 million board feet in the next year. The region's lumbermen confidently reported that up to four billion board feet of Sitka spruce stood ready "on the stump." When

Allied commanders saw the dense stands of towering Sitka spruces, they were ecstatic. A British military attaché reportedly exclaimed, "We've discovered a hidden army of giants, each one ready to rise into the skies over France." They returned to Europe buoyed by visions of unlimited spruce supply, while the sawmills eagerly got to work, and optimism radiated from both sides of the Atlantic.

Sawmills soon discovered unforeseen complications with Sitka spruce production. Only the lower trunk portion, accounting for 10 percent of each felled tree, provided sufficiently straight-grained wood for aviation. The top fifty to one hundred feet of each tree was too limby and knotty to be useful, making the original estimates of available timber on the stump seem grossly exaggerated. The painstaking milling technique also required careful hand-turning on the saw carriage to achieve precise quarter-sawn lumber, a process that could flummox even the most experienced sawyers. After a summer of frustrating work, airplane spruce production stalled at just three hundred thousand board feet, disastrously short of Allied expectations.

Meanwhile, lumberjacks endured conditions that devastated morale. Constant rainfall turned logging camps into muddy quagmires where clothing never fully dried, boots rotted off feet, and tents leaked incessantly. Primitive sanitation, little more than trenches dug into saturated earth, allowed diseases like dysentery, influenza, and pneumonia to spread unchecked. Trench foot became increasingly common, compounded by injuries from slippery logs and axes. Frustrated by these intolerable conditions, the lumberjacks organized through the Lumber Workers' Industrial Union in Spokane, Washington, threatening mass strikes unless sawmill owners met their demands for higher wages, safer working environments, and an eight-hour workday with overtime pay. When mill owners ignored these grievances, lumberjacks walked off the job, and spruce production ground to a halt.

This resulted in a rare combination of frustrated parties. European allies were furious, American military leaders were humiliated, and the sawmill owners watched helplessly as their golden opportunity for wartime profits slipped away. Nobody had any intention of compromising, and both sides turned the strike into a political showcase. The impasse demanded intervention from the one person capable of negotiating a compromise: President Woodrow Wilson.

Declaring Sitka spruce vital to national security, President Wilson directed General John Pershing, commander of the American Expeditionary Force, to create the Spruce Production Division (SPD) under the US Army Signal Corps. Pershing appointed Colonel Brice Pursell Disque, a charismatic former cavalry officer known for decisive leadership and an unorthodox approach to shipping logistics, to command the SPD and "get the spruce out." Disque immediately organized the Loyal Legion of Loggers and Lumbermen (LLLL) to build emergency railroads and erect the largest sawmill ever conceived, at Vancouver Barracks, Washington. This unprecedented military takeover of a civilian industry would become the biggest industrial undertaking the world had yet seen, surpassing even the Panama Canal construction.

Disque combed the ranks of enlisted soldiers for experienced loggers and mill workers, and compelled interested men to report for physical fitness evaluations at Army recruitment centers. Soldiers initially resisted the SPD, as logging lacked military glamour and Northwest winters were notoriously harsh. But thousands eventually showed up, motivated less by logging than by a desire for noncombat roles far from the trenches in Europe.

Within months of forming the Spruce Production Division, Disque had mobilized over ten thousand soldiers to staff sixty-two saw and axe squadrons in coastal Oregon and Washington. Once they arrived at Vancouver Barracks, Disque brought in retired lumberjacks and forestry professors from the University of Washington to give them a basic primer on how to fell fifty-ton trees without getting killed. Wearing military uniforms and wielding axes and saws, wedges and mauls, soldiers marched into the woods to work alongside roughneck civilian lumberjacks, who soon valued their camaraderie and professionalism.

"Big, husky, two-fisted fellows, young and with plenty of both muscle and brains," wrote *The Timberman*, an Oregon monthly, "they fit into the woods as a springboard fits into its notch in a tree."

Their days stretched from dawn until dusk, filled with backbreaking labor as they felled trees, stripped bark, hauled logs with winches and cables, and maneuvered through treacherous terrain littered with stumps and debris. Injuries were frequent, and vigilance was crucial for survival. But Disque's "Progressivism in khaki," as historian Harold Hyman called it, also introduced previously unthinkable comforts to logging camps: heated barracks, showers, private toilets, gyms, pianos, even motion-picture projectors for nightly screenings of silent films featuring Charlie Chaplin or Mary Pickford. These luxuries elevated morale and fostered community among men accustomed to hardship.

Private Arthur Newby with the 430th Spruce Squadron at Camp B, Snoqualmie Falls, Washington, described the conditions: "We have an excellent camp here, which is clean and sanitary. We live in bunk houses built on car trucks, about sixty feet long and divided into three rooms each. There are ten men to each room. We have all the modern luxuries—steam heat, electric lights, hot and cold water . . . There are about fifty boys here, a fine bunch, and they sure are cutting timber like real loggers."

The combination of rigorous training, modern comforts, and shared purpose transformed inexperienced young soldiers into disciplined logging crews capable of meeting, and exceeding, the ambitious demands of wartime spruce production.

With the labor situation stabilized, Disque introduced motor vehicles and built wooden plank roads to expedite spruce transport to waiting railways. His aggressive negotiations persuaded transcontinental railroads to elevate airplane spruce to priority cargo status, cutting transit times from Pacific forests to Atlantic ports from fifty days down to ten. To the delight of European aircraft manufacturers, the SPD felled every mature Sitka spruce within twenty miles of a railway, stripping the Oregon and Washington coastline bare of their centuries-old giants.

The more ambitious element of Disque's plan required building a centralized sawmill to ensure the lumber met quality standards set by the Europeans. Constructing such a facility would ordinarily take at least a year, but with SPD labor, it was built in forty-five days—just in time to announce that Disque had established an eight-hour work-day with additional overtime pay. President Wilson sent Disque a tele-gram: "I am sincerely glad to hear of the action of the lumbermen in instituting the eight-hour day. I think they will find this an act not only of wise policy but of good sense from every point of view."

Spruce timber became the backbone of Allied wartime infra-structure, used not only for airplane frames, but also for rifle stocks, ammunition crates, and cantonments lining the Western Front. Spruce railway cars carried thousands of soldiers directly into combat, high-lighting timber's essential role in wartime logistics. The trenches themselves, occupying a 475-mile-long labyrinth from the North Sea to Switzerland, were also built with spruce. Duckboards raised sol-diers above ankle-deep mud, while spruce planks reinforced trench walls and supported artillery platforms. Millions of wooden stakes anchored barbed-wire barriers in the perilous no-man's-land between opposing forces. As timber shortages in Europe intensified, Ameri-ca's 10th Engineers (Forestry) dispatched twenty thousand additional soldiers overseas to harvest European forests, maintaining the Allied forces' lumber advantage throughout the war.

The SPD had produced enough spruce to build ten thousand warplanes—six times more than Germany. These aircraft took to the skies and shifted battlefield momentum decisively toward Allied victory. British officers hailed SPD's spruce production as vital: "We could have fallen down anywhere else and still fight, but without your spruce, we'd have lost the war." Disque's program marked an unprecedented marriage of military precision, industrial ingenuity, and forestry management, exemplifying American resourcefulness at its wartime zenith.

Then on November 11, 1918, Disque received news of armistice. The war was over. The next day, he ordered all 234 spruce logging camps to cease operations. "Let every man remember always," Disque said, "that without the airplane lumber, which we, practically alone, have furnished for the entire Allied cause, this day of Thanksgiving could not have been."

The thunderous crashing of Sitka spruces gave way to subdued silence as soldiers dismantled camps and railways, decommissioned the sawmill, and sold off their equipment. After just one transformative year, the men headed home.

As the airplane spruce program faded into history, its legacies persisted. The members of the LLLL returned to their casual pursuit of timber to meet civilian demands and retained their rights to an eight-hour workday with overtime pay—two pillars of American labor policy ever since. The humble Sitka spruce—once overlooked, later revered—continued supporting America's ingenuity. Paper mills pulped it, furniture makers carved it, and Steinway & Sons crafted it into piano soundboards.

Evergreens, America had discovered, could do far more than burn, shelter, or adorn. They could also fight, fly, and win.

17

Plywood Nation

▲ ▲ ▲

AMERICA EMERGED FROM World War I with a clear lesson: Forests, if strategically managed, could decisively influence global diplomatic alliances. As peace settled uneasily across the globe, one evergreen conifer waited quietly on the sidelines, poised to step into a starring role. This mighty tree would soon revolutionize American industry, redefine home building, and equip a second war even more ferocious than the last.

In 1826, Scottish botanist David Douglas journeyed to Oregon, guided by members of the Umpqua tribe. They brought him to see a tree whose immensity left him stunned. He described it as "one of the most striking and truly graceful objects in nature," standing over three hundred feet tall with the first hundred feet completely branch-free, arrow-straight, and uniformly thick. Douglas then uttered perhaps the greatest understatement in botanical history: "The wood may be found very useful for a variety of domestic purposes."

Centuries before Douglas arrived in the Pacific Northwest, Indigenous peoples had already cultivated an intimate relationship with this remarkable evergreen. Inland tribes carved its young saplings into hunting bows and curved snowshoe frames, while Salish communities on the Pacific coast crafted an ingenious array of fishing implements from the wood and used its resin as a torch fuel for night fishing expeditions. The Bella Coola of British Columbia mixed the tree's resin with dogfish oil, applying the ointment medicinally. Other Indigenous groups valued the tree's Vitamin C–rich shoot tips as a breath freshener and sweetened their foods with melezitose, a rare trisaccharide sugar that appeared as rice-like grains encrusting the tree's branches in summertime. The Makah and Nuu-chah-nulth tribes carved the wood into sturdy, ocean-going dugout canoes, some reaching fifty feet in length, expertly navigating treacherous coastal waters on whale hunts. Native Hawaiians, more than twenty-five hundred miles away, used logs of this resilient species that drifted ashore to construct *wa'a kaulua*, their renowned double-hulled voyaging canoes. This long-standing Indigenous craftsmanship testified to the wood's durability and buoyancy, qualities Europeans would later exploit in their own shipbuilding efforts.

Coast Salish peoples knew the tree as *lá:yelhp* in the Halkomelem language and *čəbidac* in Lushootseed, but after Douglas's expedition, the scientific community struggled to agree on a proper name. Its downward-hanging cones contradicted the upward-facing cones typical of true firs, while its needles and bud forms defied categorization among known spruces and pines. Its genetic complexity (thirteen pairs of chromosomes compared to twelve pairs or fewer in other conifers) endowed the tree with unrivaled adaptability, enabling it to spread vigorously across a range of inhospitable terrain from Mexico's arid volcanic landscapes to Alaska's rain-soaked shorelines.

Botanists wrestled with how to classify a tree displaying such bewildering diversity. At various points, twenty-eight common names and seventeen scientific names were proposed, including Oregon pine, Columbian pine, false hemlock, red fir, Douglas spruce, and Puget Sound pine. Eventually, the US Forest Service settled on Douglas-fir (*Pseudotsuga menziesii*), deliberately employing a hyphenated name as a wry acknowledgment that the tree was neither Douglas's original discovery nor a true fir.

Douglas-fir became the undisputed king of the building trades, prized for its exceptional modulus of elasticity, making it the stiffest of all evergreen conifers. Its insulating properties are five times more effective than concrete, and its dimensional stability allows builders to nail it into place "green" without significant shrinkage or warping. As former *Seattle Times* reporter William Dietrich later observed, Douglas-fir occupied a unique position as "the sacred center of forest capitalism, the money tree, the sawmill dream."

The Douglas-fir's greatest contribution, however, still lay ahead. During the interwar period, technological innovation would transform this versatile timber into something revolutionary, the world's first engineered wood product: plywood.

In the late eighteenth century, Samuel Bentham discovered that by gluing together thin sheets of wood, called veneers, he could create a composite timber stronger and more resistant to warping than planks cut directly from logs. He patented his veneer-gluing technique in 1797, but his process relied on painstakingly hand-cut veneers, limiting the size and commercial viability of the finished product. While examples of veneers date back to Egyptian tombs around 2500 BC, they remained elusive as a mass-produced industrial material. Bentham intended his laminated veneer process for shipbuilding, envisioning warships that would resist the damp maritime climates and tropical humidity that weakened conventional wooden hulls. Though his naval application stalled, the foundational idea paved the way for plywood's eventual prominence.

Swedish engineer Immanuel Nobel—father of Alfred Nobel, inventor of dynamite—advanced Bentham's concept in 1850 by patenting machinery that spun logs on a rotary lathe. A precisely angled blade peeled away continuous, paper-thin sheets of veneer from the spinning log, like unraveling paper towels from a roll. Engineers could produce wide sheets from relatively slender logs, provided the timber had reasonably straight grain and minimal knots. Veneers were then glued together at alternating right angles and pressed flat under heat. But early glues dissolved in water, causing the pressed wood sheets to delaminate, relegating the invention once more to commercial obscurity for decades.

A critical breakthrough came in 1923, when the chemist Frits

Pollack developed water-resistant urea-formaldehyde resin, finally enabling veneered wood to withstand moisture and humidity without separating. Combining Nobel's rotary veneer-cutting process and Pollack's adhesive, Douglas-fir plywood gained unique dimensional stability that allowed it to be easily shaped, heated, steamed, or permanently bent. As the modern wood-products industry took shape in the early twentieth century, manufacturers embraced the term "plywood" to describe this versatile innovation. Layering veneers of exotic, decorative woods enhanced plywood for cabinet doors, cupboards, and tabletops. Its lightweight strength proved ideal for boat and aircraft construction, while its large, stable sheets provided an excellent material for sheathing floors, walls, roofs, and automobile running boards. The Douglas Fir Plywood Association described its product in a sales brochure as "a traditional material in modern form—stronger, lighter, more serviceable, and more useful" than ordinary timber, confidently predicting its "startling growth" in years to come.

Startling, indeed. By 1940, plywood had achieved widespread commercial use, revolutionizing home construction, furniture-making, and transportation. In sawmills across the Pacific Northwest, teams of workers operated rotary lathes day and night, producing veneers at unprecedented speed. Manufacturers developed assembly lines reminiscent of Henry Ford's automobile plants, where workers spread glue, layered veneers, and fused plywood panels in heated hydraulic presses. Entire towns sprang up around plywood factories, attracting skilled workers who proudly called themselves "veneer men," emblematic of America's newfound industrial efficiency. Billions of board feet of Douglas-fir logs spun through veneer lathes, stabilizing America's economy and setting the stage for another great conflict: World War II.

As war mobilization intensified, plywood became a critical commodity. This new material brought together America's finest designers, artists, and architects—not on battlefields, but hidden in drawing rooms, laboratories, and assembly lines, all creating new plywood products to support the cause. Designers Charles and Ray Eames lent their talents by reimagining standard-issue metal splints as molded-plywood versions, light enough to lift with one finger and transparent to X-rays. Over 150,000 soldiers' broken limbs healed beneath the elegantly sculpted plywood Eames splints.

When the government halted piano production to preserve strategic metal reserves, Steinway & Sons adapted seamlessly by manufacturing plywood components for wooden troop-transport gliders, which were intentionally built as disposable one-way aircraft. Pilots developed a grudging respect for the gliders' fragile plywood frames, marveling at their silent descent into enemy territory under cover of darkness.

The glider program had roots decades earlier in the career of William Boeing, a young timber merchant who arrived in Seattle in 1908 to establish the Greenwood Timber Company and expand his logging empire. Having already made a fortune cutting pines in Minnesota's North Woods, Boeing recognized the revolutionary potential of combining lightweight Sitka spruce with aluminum in aviation manufacturing. He built his first prototype airplane from this pioneering blend and founded the Boeing Airplane Company on the eve of World War I—a venture that, two decades later, led America's production of wooden troop-transport gliders in World War II. By war's end, Boeing had collaborated with Steinway & Sons to produce more than fourteen thousand gliders, including those iconic aircraft that carried British paratroopers into Normandy on D-Day.

Strapped for resources and running dangerously low on metal, Britain turned to plywood for the radical new DeHavilland Mosquito warplane, affectionately dubbed the Mossie or the Timber Terror. Constructed primarily by skilled furniture and cabinet makers in small workshops across the English countryside, Mossies required careful layering of Douglas-fir veneers bonded with newly developed synthetic adhesives, creating an airframe that was remarkably light and resilient. A total of 7,781 Mossies were built, and their success astonished pilots who were accustomed to metal aircraft. Royal Air Force crews started calling the plane the Wooden Wonder, praising its agility and speed, which frequently outpaced even the newest German fighters. This improbable wooden marvel could carry bomb loads up to four thousand pounds at speeds approaching 450 miles per hour, helping Britain reconquer the skies over Europe with its proud flying trees.

At the war's apex, plywood ascended to near-mythical status when the US military declared it an "essential war material." This marked a profound transformation: One ancient evergreen had been reshaped and reimagined to form the backbone of the world's mightiest military machine. Lumberjacks felled Douglas-fir trees by the millions; plywood barracks sprang up quickly across sprawling military bases; gritty Air Force pilots glided silently behind enemy lines in plywood troop transports; Army engineers stormed the banks of the Rhine River on plywood assault boats; the Seabees construction battalions of the US Navy erected plywood outposts throughout the South Pacific; and plywood lifeboats safeguarded vulnerable supply convoys through treacherous seas.

Navy officers, including a young lieutenant named John F. Kennedy, navigated perilous Pacific waters aboard plywood patrol torpedo boats. Kennedy's command, *PT-109*, like most PT boats, featured a lightweight plywood hull chosen specifically for its maneuverability when evading enemy fire. After the Japanese destroyer *Amagiri* rammed *PT-109* in 1943, Kennedy and his men clung desperately to the slowly sinking bow section through the night, hoping for rescue. By morning, however, as the boat began to capsize, they swam three and a half miles to an uninhabited island using plywood scraps salvaged from the wreckage as life rafts. Kennedy later credited the plywood hull's buoyancy and durability with keeping sufficient wreckage afloat during that critical first night, enabling survival for himself and most of his crew. By saving the life of the man who would become the thirty-fifth president of the United States, a humble building material crafted from evergreens proved crucial not only to the war effort but also to extraordinary acts of bravery that forever changed the course of history.

The US military's comprehensive wartime mobilization brought an industrial vision to herculean fruition. This was an engineer's dream, a forester's dream, and above all, a twentieth-century American dream: the timber industry at its most titanic and world-altering, with one heroic evergreen at its core.

18

Trial by Fire

▲ ▲ ▲

IN THE ANXIOUS aftermath of the Pearl Harbor attack in December of 1941, US military leaders feared Japan might drop incendiary bombs onto America's conifer forests. With so many able-bodied men deployed overseas, the country would be ill-equipped to fight the ensuing fires. "You are dead right about the danger of forest fires on the Pacific Coast," President Franklin D. Roosevelt warned. "It is obvious that many of them will be deliberately set on fire if the [Japanese] attack there."

To combat this imagined threat, American leaders concluded with peculiar optimism that a half-clothed teddy bear offered their best defense against hostile geopolitical firebombs. The US Forest Service introduced Smokey Bear, a shovel-wielding mascot whose booming voice echoed from radios and television screens, repeating the immortal slogan, introduced in 1947, "Only YOU can prevent forest fires!" Although Japan never weaponized wildfire, Smokey Bear's admonition struck a patriotic chord, and he became America's most beloved forest guardian. A national Advertising Council survey in 1960 ranked him second only to Santa Claus in popularity. His jocular campaign of fire prevention, bolstered by comic books, school posters, and Smokey Bear dolls mailed to children nationwide, shaped perceptions of wildfire long after World War II ended.

Smokey Bear accomplished his goals with such resounding success that it became nearly impossible to find any trace of wildfires in American forests during the thirty-five-year-long postwar housing boom.

The legacy of total fire suppression—rooted in the government's systematic displacement of Indigenous communities and its deliberate erasure of their traditional fire-management practices in the early 1900s—had in subsequent decades transformed American woodlands into vast tinderboxes congested with flammable undergrowth. Then in the 1960s and 1970s, landmark environmental laws, including the Wilderness Act and the Clean Air Act, failed to acknowledge fire's ecological inevitability, further embedding in the national psyche the misguided belief that fire was entirely unnatural and preventable. Smokey Bear's cultural triumph thus led to a dangerous ecological paradox. Protecting forests from fire only increased their vulnerability, particularly in areas where housing developments bordered highly combustible evergreen stands.

The postwar period brought fundamental changes to American domestic life—not in how homes were built, but in where they stood. Between 1946 and 1975, Americans constructed more than forty million new houses in sprawling suburban subdivisions. An entire generation left city centers for quiet streets lined with rows of boxy houses and tightly clipped lawns expressing the suburban ideal of middle-class stability. Returning GIs, whose families would become known as the baby boomers, moved to the suburbs too, reshaping the landscape and the cultural imagination as vividly as Detroit's chrome-plated automobiles and Hollywood's Technicolor sitcoms.

Securing a home, garage, and fenced yard became a nearly sacred American dream, distinctly separate from urban congestion and wilderness alike. These explicit boundaries between city and country, work and home, formed a new incarnation of wood-built living that defined itself in opposition to an earlier ideal: the restless settler-colonist utopia always keeping one step ahead of history.

Yet suburban life was far from universally accessible. Developers designed suburbs into virtually all-White enclaves, actively excluding Black people, Indigenous people, and other people of color through discriminatory practices embedded deep in federal housing policies.

Racial covenants encouraged by Federal Housing Administration guidelines, along with discriminatory lending by banks, systematically denied homeownership opportunities to minorities. As a result, Black migrants from the rural South found themselves confined to older central-city neighborhoods, creating a stark racial divide described by sociologist Reynolds Farley as "chocolate cities and vanilla suburbs." The practice of redlining—refusing mortgages to residents of Black or racially diverse neighborhoods—further guaranteed the deterioration of urban communities. Although the 1968 Fair Housing Act officially banned housing discrimination, meaningful enforcement lagged until the Equal Credit Opportunity Act of 1974 and the Community Reinvestment Act of 1977 outlawed redlining. But by then, the racial patterns had solidified. In 1980, 44 percent of Americans (around one hundred million people) lived in suburbs, 90 percent of whom were White.

While White Americans prospered in suburban enclaves, Indigenous communities faced a sharply different reality, marked by ongoing economic disadvantages and threats to tribal sovereignty. In Wisconsin, the Menominee Nation had sustainably managed its reservation forests since opening their own sawmill in 1908, employing hundreds of Native workers and saving over $10 million by 1950. Despite their relatively stable economic position, the Menominee still lagged behind White-owned industries in the Lake States. Following a prolonged legal battle over lands seized during the Dawes Act—lands that had enriched timber barons like Ezra Cornell and Frederick Weyerhaeuser—the US Court of Claims awarded the Menominee over $8 million in damages.

Yet Senator Arthur Watkins, chair of the Senate Subcommittee on Indian Affairs, attached a punishing condition to this settlement. The Menominee had to accept "Indian termination," a policy explicitly aimed at dismantling tribal sovereignty and forcibly assimilating Native Americans into White society. Watkins, a devout Mormon, insisted that termination would be "better" for the Menominee, declaring, "The time has come for us to . . . help the Indians stand on their own two feet and become a white and delightsome people as the Book of Mormon prophesied they would become."

Termination devastated the Menominee virtually overnight. Federal support for healthcare, sanitation, and fire protection abruptly vanished, leading the tribal hospital to close its doors in 1961, and forcing families to travel up to eighty miles for basic medical services. Schools shut down, unemployment soared past 50 percent, and desperate families resorted to hunting deer to stave off hunger—acts which ironically led state officials to prosecute Indigenous hunters for "poaching" on their ancestral lands. Wisconsin imposed property taxes that drained tribal finances. State-appointed overseers compelled the Menominee sawmill to increase logging beyond sustainable limits, depleting both forests and financial reserves. Eventually, the government authorized the sale of prime reservation lands to cover mounting debts, subdividing sacred tribal territory into parcels for summer cottages belonging to wealthy White families.

After decades of tireless advocacy, the Menominee successfully lobbied Congress to reverse termination. In 1973, President Richard Nixon signed the Menominee Restoration Act, reinstating federal recognition and tribal authority over their resources. But the damage was already deeply entrenched: Half of Wisconsin's Native American population had died prematurely due to inadequate healthcare and impoverished living conditions.

Even as Indigenous communities struggled with the legacies of dispossession and displacement, White Americans continued to transform the continent in distinctly different ways. Inspired by Aldo Leopold's conservationist writings and the spirit of the first Earth Day in 1970, about a million people migrated into the western forests, launching a "back-to-the-land" movement that created hundreds of new towns within an expanded suburban frontier known as the wildland-urban interface (WUI). Unlike traditional suburbs surrounding established cities, the WUI blurred boundaries, placing human development

directly against wildland vegetation. Real estate developers ignored—
or perhaps remained unaware of—historical wildfire patterns, and
constructed main streets, movie theaters, and car dealerships immediately beside dense, fire-prone woodlands. Wildfire was treated as
an unthinkable anomaly rather than the ecological inevitability it has
always been.

The balloon-framed homes filling suburban and WUI communities typically stood one or two stories tall and were built according
to one of three fashionable designs: ranch, split-level, and the ironically named colonial, imitating traditional post-and-beam structures.
These nearly identical houses differed only by subtle variations in
trim, color, and window placement, while interior details like flooring and cabinetry allowed customization for consumer satisfaction.
Every detail required significant quantities of wood, and on average,
thirty-five mature evergreens were cut down to build a single suburban
home. Whether their façades featured stucco in the West, brick in the
South, or clapboard in the Northeast, each contained an underlying
balloon-framed structure built from dimensional lumber—primarily
Douglas-fir.

"Doug fir," as builders fondly called it, had risen to this lofty position as the premier home-building tree partly due to its abundance
in easily accessible lowland valleys in the Pacific Northwest in the
early twentieth century, a time when logging had already depleted
the nation's eastern pine reserves. Building upon the Army's master-
fully coordinated harvest of Sitka spruce during World War I and the
widespread adoption of Douglas-fir plywood during World War II,
the logging industry ushered in a new era of rapid evolution during the
postwar housing boom. Methods for dismantling trees shifted decisively from manpower to machinery, and commercial uses for wood
products grew exponentially.

Pacific Northwest loggers began to tackle giant Douglas-firs and
western red cedars (*Thuja plicata*) on steep slopes, pioneering innovations that would become mainstays of modern forestry. Diesel engines,

bulldozers, and especially high-lead cable rigging systems—deploying aerial pulleys mounted atop living trees—allowed loggers to extract timber from previously unreachable mountainsides and deep ravines. Timber companies recruited their first high riggers from the maritime industry, where sailors were already accustomed to perilous climbs up ships' masts. Equipped with three-inch climbing spikes and carrying saws and axes, high riggers tossed heavy ropes around tree trunks and their own waists, rapidly ascending two-hundred-foot-tall evergreens. At dizzying heights, they axed off top leaders as wide as mature New England pines, clinging to trunks that whipped violently beneath them.

Ground-based donkey engines, anchored to multiple spar trees, then pulled felled logs uphill using elaborate spiderwebs of steel cables stretching hundreds of feet across valleys. The work could turn deadly, fast. Snapped cables, equipment failures, and falling timber claimed hundreds of lives annually. Between 1945 and 1970, logging surpassed mining to become America's most dangerous occupation, with fatality rates twice that of coal mining and thirty times greater than manufacturing.

The success of logging operations depended on the high rigger's skill in placing cable systems, which officially ended the era of dragging logs with oxen and horses. Over the decades, Pacific Northwest valleys were scoured and scarred, littered with rusted machinery and abandoned cables. Though less wasteful than earlier lumbermen, these loggers still discarded vast amounts of unmerchantable wood in slash piles. The forests around Puget Sound were steadily leveled to supply booming suburban and WUI construction, making Oregon the leading softwood lumber-producing state—a title it never relinquished.

By 1960, gasoline-powered chain saws had replaced axes. Cantankerous, deafeningly loud, and spewing black exhaust, early chain saws weighed over a hundred pounds and featured intimidating six-foot blades studded with hundreds of sharp teeth that were modeled after the C-shaped jaws of timber beetle larvae. The teeth cleanly removed wood chips with each revolution of the blade, allowing chain saws to slice through thick trunks like a warm knife through butter.

Chain saws became both a thrilling symbol of humanity's dominion over nature and a devastatingly efficient industrial tool that changed logging forever. By requiring less brute physical strength to operate than traditional manual tools like axes and misery whips, chain saws helped open the forestry profession to greater participation by women. Complementary innovations like log skidders, excavators, and all-terrain tracker vehicles shrank colossal trees down to human scale, allowing virtually anyone to fell giants with remarkable ease.

By the 1980s, swashbuckling loggers could be seen rappelling down sheer cliffs, cutting trees and calling in helicopters costing $5,000 per hour to airlift their quarry. Chain saws evolved into sleek fifteen-pound marvels with piston ring cutting blades spinning at eighty-eight feet per second, slicing through timber with ferocious speed. Perhaps the most impressive advance in a generation, terrifyingly efficient feller bunchers used whirling cutting heads on serpentine crane arms to grasp and cut trees into computer-guided lengths. The lumberjack of this modern era—if one could still be called that—now sat comfortably in a cushioned leather seat inside a climate-controlled cabin, manipulating a joystick to fell trees in seconds. As loggers' wildest dreams materialized through roaring machinery, evergreens that had stood for centuries were reduced to neatly sawn planks in hours.

The mechanized conquest of America's forests facilitated the rapid clearing of its last truly ancient trees, marking a transformation as profound as it was irreversible. In the great Douglas-fir and western red cedar stands along the continent's westernmost edges, big-tree logging faced an inevitable reckoning. There was no frontier left to chase. The industry would have to sustain itself on whatever remained.

By 1990, the nation's relentless demand for dimensional lumber had reshaped the landscape, literally, by building a new frontier in sprawling suburban and WUI communities, where more than 60

percent of Americans now lived. Small towns and once-sleepy farming settlements morphed into bustling suburban centers. Douglas-fir
dimensional lumber became so ubiquitous that builders popularized a saying: "If Doug fir disappeared, America itself would vanish
overnight." This frontier was no longer defined by geography, but by
evergreen lumber's pervasive cultural, economic, and architectural
influence on the American way of life.

Suburbs and WUI towns, as author John Vaillant described in
the book *Fire Weather*, represented an "eighty-year jackpot rollover
of unburned trees, and its half-century bonanza of vinyl-sided, tar-
shingled plywood houses." Steeped in highly flammable resins derived
from evergreens, these homes stood as precarious monuments to
America's complex relationship with wood and wilderness, awaiting
fire to reclaim its ancient dominion.

Climate change ushered in rising temperatures and prolonged
drought, setting the stage for what ecologists termed a "perfect storm"
of vulnerability. A US Geological Survey report revealed that drought
stress in western forests during the late twentieth century elevated tree
mortality rates tenfold above historic norms. As trees weakened, ravenous beetles—each no bigger than a grain of rice—bored through the
bark of spruce, pine, and fir trees and laid eggs beneath the surface.
Mild winters allowed larvae to hatch in unprecedented numbers, and
these young beetles feasted on the trees' soft inner tissues, cutting off
essential nutrients and starving their hosts from within. Under normal
conditions, conifers would defend themselves by drowning invading
beetles in sticky resin, but drought-stressed trees could no longer produce enough protective sap. The beetle epidemic ultimately devastated
eighty-eight million acres of western forests, leaving mountainsides
carpeted with rust-colored needles known among firefighters as "red
zones." More than 165 million dead spruce, pine, and fir trees stood
like dry matchsticks awaiting ignition.

While definitions of "destructive" may vary, according to the
National Interagency Fire Center's data logs, twelve of the fifteen

most destructive wildfires in American history occurred after 1980, with each subsequent year bringing seven more major fires and an additional ninety thousand scorched acres than the year before. The largest wildfire of the 1980s burned eight hundred thousand acres in Yellowstone National Park, defying the best efforts of four thousand US military personnel dispatched to control it. Other significant fires soon followed. The 1991 Tunnel Fire in the WUI hillsides of Oakland, California, destroyed more than three thousand homes and claimed twenty-five lives, as 1000°F blizzards vaporized entire neighborhoods.

During the first quarter of the twenty-first century, at least eighteen wildfires in the United States caused more than $1 billion in damages each—including the 2018 Camp Fire, which obliterated the town of Paradise, California, in just four hours, killing eighty-five residents and incurring at least $12 billion in losses. In Alaska, the 2019 wildfire season torched two and a half million acres, while in Montana in 2021, wildfire consumed three-quarters of a million acres. And in 2025, despite billions spent battling flames, the costliest natural disaster in US history engulfed the affluent Los Angeles County neighborhoods of Altadena, Pacific Palisades, and Malibu. Hundred-mile-per-hour Santa Ana winds sent flames tearing through streets, causing at least twenty-nine deaths, destroying sixteen thousand structures, and resulting in economic losses estimated at $275 billion.

Wherever these fires erupted, their damage extended far beyond immediate burn zones, affecting human health and disrupting communities for years afterward. Air quality deteriorated to levels exceeding hazardous thresholds, compelling hospitals to treat thousands suffering from severe respiratory illnesses. Entire cities were forced indoors, with elders recalling "ash rain" reminiscent of the 1980 Mount Saint Helens eruption. Charred soil became hydrophobic, repelling water and triggering flash floods that sent torrents of mud and debris through burned-out skeletons of homes. Economically, wildfires reduced private capital and damaged essential public infrastructure, forcing business closures and ultimately inflating the cost of living and insurance premiums in regions repeatedly devastated by uncontrolled infernos.

As the climate continues warming, wildfires will remain an ever-present threat capable of transforming remote evergreen forests and densely populated communities alike into self-fueling blowtorches.

Although modern megafires represent a comparatively modest threat in forests east of the Mississippi, decades of fire suppression there have caused a subtle but equally destructive ecological shift known as mesophication. Evergreens like the eastern white pine and the Atlantic white cedar evolved to survive periodic fires, but seedlings cannot regenerate beneath dense, closed canopies left unburned for over a century. Without fire to open the canopies and allow light in, sun-loving conifers gradually succumbed to shade-tolerant broadleaf hardwoods like red maple. This slow-moving phenomenon pushed forests toward a "mesophytic abyss," a self-reinforcing cycle in which shade perpetually begets more shade, permanently altering forest composition. Compounding this crisis, eastern white pines suffered from a fungal pathogen called Caliciopsis canker, causing needle loss and resinous lesions, weakening trees further and accelerating branch dieback.

Simultaneously, invasive woolly adelgids—aphid-like insects inadvertently imported from Japan on nursery stock—ravaged eastern hemlocks. Historically prized by the leather industry for their tannin-rich bark, these evergreen conifers suffered widespread losses as the adelgids spread through the Appalachians starting in the 1990s. Achieving kill rates of 80 percent, the infestation threatened over 170 million hemlocks in Michigan alone. Entire populations vanished in Great Smoky Mountains National Park, where hemlocks once shaded creeks and sustained the cool microclimates essential for native trout.

An evergreen forest can mask its collapse for a deceptively long time through adaptive responses such as those seen in California's giant sequoias or the South's longleaf pines, which shed older needles during drought. This selective foliage loss reduces water demands, allowing trees to maintain a healthy-looking green crown despite their internal stress. In the case of giant sequoias, areas experiencing high-severity fires compounded by drought conditions have seen over 90 percent of mature trees perish, with insufficient seedling regeneration to replace them. This underscores how the outward resilience of evergreen forests can obscure the underlying vulnerabilities and the potential for sudden, large-scale ecological shifts.

But now, with the nation's evergreen forests no longer able to conceal their decline, satellite imagery has laid bare a continent in distress. Since 1972, NASA and the US Geological Survey have deployed satellites that snap specialized digital photographs of the earth's surface, allowing scientists to track ecosystem-level disruptions that are difficult to see from the ground. Western US landscapes appear as a patchwork quilt of black burn scars, thin green valleys, and brown expanses representing heat-stressed, beetle-ravaged evergreens. Satellite images of eastern landscapes reveal dark patches of dead hemlocks, yellow blotches of dying pines, and light-green areas where mesophytic maples have taken hold. As evergreens turn ever-brown, the continent itself appears afflicted by a virulent skin disease.

Recognizing the severity of the issue, the Forest Service released its Wildfire Crisis Strategy in 2022, calling for controlled burns and replanting on twenty million acres of public forests and another thirty million acres on state and private lands. Despite spending $100 million on reforestation initiatives that year, the agency treated only 6 percent of the acreage burned in the previous decade—and this effort remains vulnerable to political whims. In 2025, federal budget cuts aimed at "government efficiency" abruptly terminated the employment of two thousand Forest Service workers directly involved in prescribed burning and replanting programs, casting further doubt on America's ability to sustain the most basic forest restoration efforts.

All of this has created a pernicious cycle. Federal, state, and local firefighting agencies remain perpetually stretched thin, lacking the resources and funding to contain new fires, let alone preempt larger infernos with smaller controlled burns. Replanting America's scorched landscapes at appropriate scale would require more than five billion conifer seedlings annually, but the nation produces just one and a half billion—and 94 percent of them are earmarked by private corporations for intensively managed loblolly pine plantations in the South. This virtual monopoly on nursery capacity has curtailed the nation's ability to restore the backlog of lands ravaged by fire and beetles, reducing the reforestation pipeline to a slow trickle. Without enough seedlings, nor workers to plant them, those charred mountainsides will likely remain ash brown for the foreseeable future.

19

Green Glacier

▲ ▲ ▲

AMERICA HAD SEEN it all before. Ambitious government interventions intended to repair or prevent environmental catastrophes often began with earnest intentions but led to unintended consequences. In the 1930s, after Dust Bowl storms stripped away precious topsoil and buried farms beneath drifts of sand, President Franklin D. Roosevelt championed an unprecedented tree-planting campaign under the New Deal. Congress allocated $14 million in emergency relief to create the Prairie States Forestry Project, an initiative designed to halt soil erosion by planting 220 million trees in shelterbelts spanning 18,600 linear miles from Texas to North Dakota. Some politicians mocked Roosevelt's tree-planting scheme as a naïve attempt to tame the stubborn prairie. South Dakota's governor, Tom Berry, called the shelterbelt initiative foolish and predicted its inevitable failure. Editorial cartoons of the era lampooned Roosevelt as an overzealous Johnny Appleseed, scattering seeds destined never to sprout. As history unfolded, their skepticism would begin to look prophetic.

Three million young men enlisted in Roosevelt's Civilian Conservation Corps (CCC), a New Deal program created to employ farmers and laborers impoverished by the Great Depression. After the Dust Bowl disaster, thirty-three thousand of them were dispatched to the Great Plains. For six-month shifts, they lived and worked in some of the twenty-six hundred isolated CCC camps scattered across remote Midwestern prairies. Their mission of planting trees in shelterbelts around farms and towns became far more than a straightforward ecological restoration effort. Workers found the prairie's hardened soil nearly impossible to penetrate with basic shovels, forcing them to use steel pickaxes before planting. At night, exhausted CCC members joked grimly about having to plant each tree twice—once to break the soil and again to secure the seedling.

Living quarters consisted of wooden barracks heated by coal stoves, where freezing winters and blistering summers tested the resolve of young recruits. Each CCC camp, housing no more than two hundred men, was its own isolated universe complete with infirmaries, mess halls, and recreation centers—oases of civilization within otherwise desolate grasslands. Camp life unfolded largely beyond public scrutiny and the conventions of family and church, providing the young tree planters with ample opportunities to forge emotional and physical bonds that upended rigid societal norms around masculinity. Bachelor weddings between two men, playful gender-bending theatrical performances, and overt expressions of male affection regularly filled the pages of the CCC's weekly newspaper, *Happy Days*. Indiana University historian Colin Johnson argues convincingly in "Camp Life: The Queer History of 'Manhood' in the Civilian Conservation Corps" that "even if the vast majority of young men who moved out of their adolescence in rural areas and into adult manhood by way of a term or two in the CCC never thought that they had been part of anything that could or should be considered part of 'queer' history, the record strongly suggests that life in the CCC was anything but straight and narrow."

The mythological ideal of rugged masculinity figured prominently in the literary imagination of the nineteenth century, most notably in James Fenimore Cooper's Leatherstocking Tales, a pentalogy of novels published between 1823 and 1841. Cooper's celebrated frontiersman, Natty Bumppo, embodied what Teddy Roosevelt later championed as the essential traits of American manhood: strength, self-reliance, and solitude in the wilderness. Roosevelt insisted that men who ventured into untamed landscapes inevitably forged "manly qualities that are invaluable to a nation." In this narrow view of life on the American frontier, lumberjacks and cowboys stood fundamentally apart from domesticity and civilization. Their masculinity, isolated in body and spirit, was defined precisely by their separation from femininity and softness.

Yet Cooper's characters subverted this stark binary. His frontiersmen formed intense, romantic friendships, most famously exemplified by Natty Bumppo's lifelong bond with Chingachgook, a noble Mohican warrior portrayed as his closest friend and spiritual brother. Chingachgook embodied both the idealized qualities Cooper ascribed to Native Americans and a profound potential for cross-cultural understanding. Literary critic D. H. Lawrence memorably characterized these relationships as "deeper than the deeps of sex." Such male bonds were implicitly homoerotic yet existed without the pejorative categorization of deviant sexuality that dominated American discourse. Thus, from the inception of American frontier mythmaking, Cooper acknowledged same-sex love, contradicting later claims that homosexuality was a recent or unnatural phenomenon.

This nuanced portrayal of intimacy between men also reflected real-life frontier experiences. Whether the lumberjacks who tackled California's giant sequoias in the 1880s, Loyal Legion recruits harvesting Sitka spruce during World War I, or the CCC workers who planted Roosevelt's shelterbelts, all participated in an American tradition of complex male intimacy forged by the crucible of hard labor. Their bonds, like those between Cooper's frontier heroes, could be romantic and even erotic, yet they were not viewed through the modern lens of abnormality. New York University professor Christopher Packard, in *Queer Cowboys*, argues that these homosocial relationships were not mere situational conveniences but vital emotional lifelines, reinforcing masculine privilege precisely through intimate dependence.

By the time of the Great Depression, however, American masculinity had hardened into rigid definitions that excluded same-sex attraction. Beneath the official narrative shaped by Roosevelt's rugged individualism and reinforced through Hollywood depictions of solitary cowboys, the authentic lived experience of American men remained richer, more complicated, and far more fluid.

In this context, the shelterbelts planted by CCC men represented more than windbreaks or erosion control; they challenged the mythologized ideal of predominantly White, rugged, and heterosexual masculinity. Though the Prairie States Forestry Project did not consciously set out with this intention, it inevitably nudged American culture toward recognizing intimacy between men as natural, commonplace, and fundamentally woven into the nation's fabric.

This nuanced historical perspective offers something profoundly meaningful that many LGBTQ people still lack today: a solid sense of legacy, a continuous and detailed presence in the American story. Just as the histories of Native Americans and African Americans are not niche narratives separable from American history—but are instead central to it—Queer history is integral to America too. Queer lumberjack history, far from being a mere footnote, is foundational to America's growth and economic prosperity.

Yet even as these societal shifts went largely unnoticed, a more visible change swept across the Great Plains. For generations, the region had been defined by its simplicity, its space, and its elemental openness. Relatively featureless grasslands were home to prairie chickens, pronghorns, and myriad insects that sustained a complex food web. Novelist Willa Cather described the Nebraska prairie of the early twentieth century as rolling endlessly beneath skies so expansive they stretched beyond comprehension. "Trees were so rare in that country, and they had to make such a hard fight to grow," Cather wrote in *My Ántonia*, "that we used to feel anxious about them, and visit them as if they were persons. It must have been the scarcity of detail in that tawny landscape that made detail so precious." The American prairie of Cather's childhood has been transformed in ways she could scarcely have imagined.

Many of the shelterbelt trees planted by CCC workers were eastern red cedars (*Juniperus virginiana*), hardy evergreen conifers native to the eastern United States and resistant to drought, heat, and cold. They thrived in a harsh biome that Indigenous peoples had long managed by setting carefully timed fires to prevent woody encroachment on the grasslands. For most of human history, trees remained confined to the prairie's deepest ravines. But when fire suppression became federal policy and Smokey Bear commanded the airwaves, cedars crept from their secluded hiding spots and advanced unchecked across the open prairie in every direction.

The heartland turned upside down. Rich grasslands and their abundant wildlife gave way to monoculture cedar stands nearly devoid of biodiversity, creating silent, shadowy spaces where prairie birds like meadowlarks and bobolinks could no longer nest or forage. The ecological shift was so profound that ranchers and ecologists named it the Green Glacier. Unlike its icy namesake, this glacier was alive, its slow and steady march measured not in ice floes, but in red cedars spreading relentlessly north and west across America's heartland. Satellite imagery revealed that the cedars' range expanded by 2 percent per year between 1970 and 2025, ultimately overtaking forty-four million acres of the Great Plains. The once-unbroken horizon has succumbed to a rising tide of evergreens as human actions intended to preserve the prairie instead push it toward the brink of extinction.

Cattle ranchers in regions like Nebraska's Sandhills or South Dakota's West River, who depended on nutritious prairie forage, watched helplessly as their pastures became crowded with cedars so dense that the ground beneath them lay dark and lifeless. Native wildflowers were smothered under cedar needles and waxy, berry-like seed cones that acidified the soil. Ranchers from the Flint Hills of Kansas to the Missouri Breaks of Montana who once watched thunderstorms roll in from forty miles away suddenly found themselves hemmed in by walls of cedars. With prime ranchland now transformed into worthless cedar thickets, cattle grazing capacity dropped by 75 percent in some areas, forcing ranchers to sell off portions of their herds or abandon ranching altogether.

Cattlemen desperate to reclaim their pasturelands faced limited options. Controlled burns offered temporary reprieve, but federal policies and public fears about fire limited their use. Mechanical removal of the cedars required heavy machinery and skilled labor, both of which were increasingly rare and expensive commodities. Those who were compelled to hire mechanical clearing crews paid upwards of $300 per acre—an economic burden that drove families off land their ancestors had homesteaded during the Anglo-American Settler Revolution.

The red cedar's advance was aided by deeper cultural attitudes toward wilderness landscapes. "Ironically, Cather's romantic vision, no matter how gripping, was rooted in the same settler-colonial mindset that ushered forth the Green Glacier," wrote Nebraska journalist Carson Vaughan in *The New York Times*. "Ultimately, both were acts of erasure."

America was, after all, a nation shaped by evergreens. Ever since the Pilgrims stepped ashore, forests provided stability, safety, and permanence. The prairie, by contrast, had always been viewed as an empty canvas by immigrant newcomers. It was a vast expanse needing improvement, a landscape perpetually incomplete. The urge to plant trees across the Great Plains echoed a centuries-old cultural impulse. Forests were civilization; trees were progress. But what began as progress eventually ended in crisis.

Ironically, at precisely the moment America most needed a robust logging infrastructure in the Midwest to remove these invasive evergreens, its modern timber industry stood gutted by decades of mechanization and economic restructuring. As timber harvesting became increasingly automated and lumberjacks were replaced with computerized feller bunchers, hydraulic cutters, and robotic sawmills, skilled labor once abundant in rural communities had all but vanished. Between 1950 and 2023, timber industry employment fell by 75 percent in the United States, leaving only 2,510 sawmill businesses in operation—a drastic drop from the estimated thirty-two thousand sawmills recorded in the 1840 census.

The decline in one of America's most historically crucial industries became particularly evident during the COVID-19 pandemic. Home buying soared as people who still had jobs and income to spare exited cities for the countryside, springing for second homes in a new remote-work economy. Workers who stayed in existing homes built new office additions or tackled long-delayed renovation projects. Record-low interest rates facilitated a frenzy of construction activity, but when unprecedented demand for lumber collided with a workforce too diminished to respond, supply chains fractured. The retail price of dimensional lumber and plywood quadrupled. Sawmills, already short-handed when demand surged, temporarily shut down to implement safety protocols, making it impossible to meet the rush. And as lumberyards ran empty, unwanted cedars continued spreading unchecked across millions of acres of the Great Plains.

Panic buying and supply-chain bottlenecks led to store shelves being emptied of modern civilization's new Achilles' heel: toilet paper. Few primal fears proved more distressing than running out of toilet paper during a global pandemic. Supermarkets instituted purchase limits, shoppers waited in snaking lines, and price gouging briefly made toilet paper more precious than gasoline. Shocking amounts of North America's ancient forests had been pulped for toilet paper in the nineteenth century, meaning society literally wiped its collective behinds with trees older than Jesus Christ. Given America's previous willingness to destroy ancient forests for disposable products, converting invasive cedars to toilet paper could have solved two crises simultaneously. Remarkably, no such effort materialized.

Despite having virtually unlimited access to the most advanced machinery in human history, the modern timber industry lacked surge capacity. "Because work in the wood products industry is becoming more mechanized, the workforce now needs to be more technically competent and highly trained," said Jeffrey Prestemon, a senior scientist at the Forest Service. "It takes times to develop these skills, so ramping up production in a short period of time is not as easy as it was in the past."

America confronted a shocking irony. For centuries, the nation had struggled to satisfy insatiable demand for evergreen timber, exhausting old-growth forests from Maine to Michigan, from North Carolina to Louisiana, and from California to Oregon. Now, the nation faced precisely the opposite problem. Here stood forty-four million acres of unwanted cedars suffocating the prairie states, awaiting a harvest that would likely never come. They represented an abundance without purpose, a resource without value. For the first and only time in its long and tangled history with evergreens, America had too many trees, and too few people willing or able to cut them down.

Cather's writings probed themes of fleeting aspirations, the disillusionment following idealistic pursuits, and the bittersweet remnants that linger after dreams are surrendered. In O Pioneers!, she described America's prairie landscape as a "great fact" that could "overwhelm the little beginnings of human society that struggled in its somber wastes."

In their relentless advance, the cedars pose a haunting question. Can America reclaim the evergreen world it has inadvertently created? Or is the prairie destined to become, as Cather once suggested, "not a country at all, but the material out of which countries are made"?

20

And the Trees Will Last Forever

▲ ▲ ▲

AMERICA HAS TRAVELED far since colonial times when forests were hastily leveled to build a nation. Over a quarter of the way through the twenty-first century, public outcry over clear-cuts and stark images of barren landscapes broadcast widely across television screens and social media helped bring an end to the era of reckless, conveyor-belt logging. An increasingly sophisticated understanding of sustainability has profoundly reshaped the timber industry. The exploitative ethos of "cut out and get out" has given way to careful, science-based management of second- and third-growth evergreens, aligning profitability with conservation and adapting forestry practices to global markets and consumer tastes. Where loggers once wielded only axes and saws, timber companies now employ entire teams of specialists—hydrologists, archaeologists, wildlife biologists, community liaisons, and lawyers—to ensure long-term forest health and regulatory compliance.

Perhaps no company embodies this evolution more visibly than Weyerhaeuser, which is now the second-largest landowner in the United States after the federal government, controlling approximately ten million acres. After facing prolonged environmental protests and consumer boycotts, notably from organizations like Rainforest Action Network, Weyerhaeuser understood that its future depended on public trust as much as timber. Consequently, the company has transitioned from an extractive logging giant into a corporate steward of evergreen forests. Recognizing that scattered natural regrowth alone would not reliably replenish harvested forests, Weyerhaeuser made deliberate replanting a cornerstone of its business model. It began systematically replacing every felled tree with seedlings nurtured in state-of-the-art company nurseries, using proprietary genetic lines of Douglas-fir selectively bred for rapid growth, disease resistance, and optimal timber yield.

Through most of modern history, widespread replanting was economically unattractive, delaying returns on investment by decades. But shifting consumer preferences toward sustainably sourced wood, combined with emerging carbon-credit markets in the early 2000s, made replanting trees fiscally prudent and strategically advantageous. Weyerhaeuser was already half a century ahead of the curve, having established Tree Farm No. 1 in 1941—a pioneering project covering 120,000 acres in western Washington. The ambitious initiative converted deforested land into a showcase for managed forestry, complete with rotational harvest cycles and preemptive fire controls. Treating forests explicitly as renewable agricultural crops, Weyerhaeuser planted, grew, harvested, and then replanted at thirty- to forty-year intervals. Compared to barren stump fields, these replanted landscapes represented clear ecological progress by providing wildlife habitats, erosion control, carbon sequestration, and improved water quality. Perhaps most importantly, intensively managed tree farms alleviated logging pressure on wild forests. By efficiently meeting society's demand for wood products, these plantations thus enabled large, ecologically complex old-growth forests to remain intact. In this sense, corporate plantations and conservation areas could coexist, demonstrating a pragmatic balance between economic interests and conservation.

Critics of managed forests often equate planted timber stands to artificial or "man-made" ecosystems. Just as a backyard swimming pool cannot replicate the complexity of a natural lake, they argued, a carefully cultivated tree farm falls short of the dynamic species interactions that characterize old-growth forests. While common wildlife such as deer and black bears benefited from increased cover provided by monocultural stands of juvenile evergreens, other species that evolved to live exclusively in old-growth forests, like the spotted owl and the marbled murrelet, saw their essential habitats diminished.

Though no single ecologist is credited with originating the swimming-pool analogy, landscape architect and environmental theorist Ian McHarg notably popularized similar ideas in his seminal 1969 work, *Design with Nature*. McHarg argued extensively that engineered landscapes, however beneficial, lack the resilience, species diversity, and adaptive qualities intrinsic to untouched ecosystems, underscoring the critical need to preserve genuinely natural habitats.

Seen in this broader context, Weyerhaeuser's business model represented a broader transformation in American forestry. It reflected and amplified a national shift toward viewing forests as renewable assets rather than expendable commodities. Maximizing shareholder value now meant careful stewardship of resources to harvest and regrow trees repeatedly over decades. Profitable forestry became synonymous with sustainability.

Weyerhaeuser's path toward sustainable forestry echoed a similar approach launched decades earlier by the Menominee Nation of Wisconsin. After White timber barons had stripped the North Woods bare, the Menominee's story hinted at a different, more hopeful possibility of a sustainable relationship with their ancestral lands. Despite having their once-vast evergreen empire reduced to a postage-stamp reservation, their forests became an island of hope amid an ocean of clear-cut ruin.

After successfully winning federal recognition from President Nixon in 1973, Menominee Tribal Enterprises began revitalizing its ancestral pinelands with remarkable ecological and economic foresight. By selectively harvesting only mature and diseased trees that typical logging operations would consider junk, it nurtured the most productive forest on the continent. Resident foresters replanted three hundred acres each year with saplings cultivated in their own nursery, where they extracted seeds from hand-harvested cones and then planted in test plots assessing climate adaptability. In the half century after regaining tribal authority over their resources, the Menominee had harvested double their forest's original timber volume yet maintained 40 percent more standing wood than when they began. Their approach baffled timber experts, who called it the "forester's paradox," a counterintuitive truth that cutting down trees could yield more robust growth and improve forest productivity. "Even some Indians want us to leave the forest alone, but we have centuries of data and know that humans interact with the forest, it affects us, and we affect it," explained Michael Skenedore, president of Menominee Tribal Enterprises. "I view standing trees as our inventory, and why would any business let its inventory rot or spoil? To cut it down is to save it."

When Wisconsin economic development officials invited a group of international wood buyers to the Menominee sawmill in 2023, lumber merchants from Mexico, Taiwan, and Thailand arrived eager to secure premium timber from one of North America's most renowned forests. Upon arrival, they were surprised to discover no immediate sales opportunities. Skenedore informed them the tribe had already sold its entire milling output for the year. "You can only place preorders now," he told the assembled buyers. "We don't have much left to sell, so you're here to learn."

On pine-cutting day, steam rose from vents, and the air inside the Menominee mill buzzed with a mechanical symphony: the shriek of saw blades, the clank of conveyor belts, and the intermittent whistles signaling timber sliding between cutting stations. Workers expertly cut logs into boards varying from modest four-inch widths to planks exceeding two feet. Thousands of logs lay stacked outside, a forest in repose, ready to be milled. Most logs would become framing studs or barn siding, while offcuts and sawdust were repurposed either as packaged animal bedding or as fuel to generate electricity. This self-sustaining practice saved the tribe approximately $20,000 per month in natural gas costs, exemplifying a sophisticated business model rooted in sustainability.

Later that evening, locals gathered at a diner on the outskirts of town for a traditional fish fry. Pickup trucks caked in white sawdust lined the parking lot. Inside, a bulletin board hung in the dimly lit hallway beside the kitchen, crowded with posters bearing haunting reminders of deeper struggles within the community:

Searching for Katelyn. $5,000 reward.
If you have any information of Rae's whereabouts,
please call the hotline.
We haven't heard from Lisa since January.

Corners of faded posters curled back, revealing older pleas for help underneath. Outside, similar posters taped to truck windows displayed photographs of missing women, each one a descendant of proud foresters from the great North Woods.

At least twenty-two Menominee women had gone missing, some for weeks, others for months or even years. Their disappearances formed part of the Wisconsin Missing and Murdered Indigenous Women Task Force's broader fight against abduction and trafficking. This local tragedy mirrors a nationwide crisis, with more than fifty-seven hundred Indigenous women reported missing, a catastrophe only now beginning to receive the public attention it urgently deserves.

Nevertheless, this tiny, hardscrabble place has signposted a viable off-ramp from the tragedy of the commons. After the Menominee's revered nineteenth-century leader, Chief Oshkosh, negotiated the treaty securing the reservation for his people in 1854, he laid down a lasting directive for intergenerational prosperity:

Start with the rising sun and work toward the setting sun, but only take the mature trees, the sick trees, and the trees that have fallen. When you reach the end of the reservation, turn and cut from the setting sun to the rising sun, and the trees will last forever.

21

Evergreens Everywhere

▲ ▲ ▲

SIXTY YEARS AFTER *A Charlie Brown Christmas* first aired, it still resonates as a critique of commercialization. The Christmas tree remains one of the last bastions of nature allowed into modern American homes, echoing Charlie Brown's original yearning for authenticity amid a holiday increasingly dominated by plastic and glitter. Although Charlie Brown's forlorn little tree may have been cut from the wild, today's evergreens bound for holiday displays rarely come from untouched forests. Instead, they are cultivated on tree farms, tended with meticulous care, and managed much like the veal calves of horticulture: fed and groomed obsessively, then harvested young.

Christmas tree farming, now a sophisticated agricultural enterprise, relies on precise science rather than simple tradition. Farmers carefully control variables such as tree spacing, nutrient inputs, and pruning schedules, all of which determine a tree's shape, needle retention, and overall marketability. Every decision, from soil amendment to harvesting method, follows detailed protocols intended to create the perfect holiday specimen: symmetrical, dense, highly aromatic, and of course, uniformly green.

In Oregon's Willamette Valley, Tyler Stone, owner of Tree Kings Oregon, runs one of America's largest and most ambitious Christmas tree businesses. Stone's father launched the family farm with money borrowed from a reluctant aunt, who agreed only after Stone's grandfather had dismissed Christmas trees as a foolish investment. "He said he knew it was a bad idea," Stone recalled, swinging an eighteen-inch razor-sharp machete through branches, shaping a Nordmann fir into a pleasing, marketable pyramid. "My collarbones are permanently floating from a lifetime doing this. I stopped surgeries years ago and just accepted it as an occupational advantage."

Stone's farm showcases the complexities of this uniquely American industry, facing overlapping challenges of rising labor costs, outdated equipment, skyrocketing fertilizer prices linked to global conflicts, and increasingly extreme climate events. In 2021, an unprecedented "heat dome" scorched the Pacific Northwest, trapping intensely hot air beneath a stationary high-pressure system for more than a week. This historic phenomenon, described by climate scientists as a once-in-a-millennium occurrence, killed millions of Christmas tree seedlings and left row after row of mature trees browned and brittle in temperatures as high as 115°F. The damage was so extensive that farmers across Oregon reported losing entire generations of trees. The economic consequences will not be fully realized until a decade later, when trees intended for harvest simply will not exist.

Tree farmers like Stone rank among agriculture's biggest gamblers, rolling the dice every time they plant a seedling, betting it can survive long enough to reach marketable size years into the future. No insurance covers the risks inherent in their trade: unpredictable weather, fluctuating consumer tastes, and marginal growing conditions. Scotch pine, once America's most popular Christmas tree, lost favor due to stiff, prickly needles and a notorious tendency for premature needle drop, leaving living room floors blanketed with debris by Christmas morning. In its place, consumers now prefer softer varieties like balsam, Fraser, noble, and Nordmann firs, whose dense foliage and strong branches better align with modern expectations.

Compounding these challenges, Christmas tree farms often occupy the poorest agricultural land on rocky hillsides and clay soils unsuitable for most crops. As one seasoned Vermont grower observed, profitable crops like marijuana and corn claim the most fertile farmland, leaving Christmas tree farmers to wrestle with the leftovers. Stone's personal concerns mirror a broader demographic trend within the industry, where the average farmer is sixty-four years old and succession remains deeply uncertain. With razor-thin margins and decade-long production cycles, the industry's future seems as fragile as a six-inch-tall seedling growing under a heat dome.

Cultivating Christmas trees is anything but simple, requiring continuous and intensive manual labor with frequent nips and tucks to maintain their picture-perfect appearance. Starting in the tree's third growing season, field crews prune off the lowest six inches of limbs to create a handle at the base of the trunk to accommodate a tree stand years down the road. Then in June, workers shape the trees with machetes, seeking perfect symmetry—a hazardous task, prone to accidents like slit boots, sliced hands, and even injured dogs. Pruning methods also must be tailored to each species. Firs require trimming back latent buds, the thimble-shaped nodes along the branches, while pines are trimmed only when new needles reach half the length of the older growth.

After the spring growth hardens in July, crews swarm through the fields using bamboo sticks and twist ties to support droopy treetops, ensuring that every tree develops the desired straight spire. Workers then bring out a special device resembling a cigar cutter to score the bark of the topmost leader, which restricts sap flow and stunts the trees' vertical growth, stimulating fuller lateral branching. Without this critical step, trees become gappy, an undesirable trait to buyers looking for lush foliage. Growers also pinch off female cones by hand, because cones divert valuable nutrients from branch growth, and consumer surveys consistently show that shoppers prefer trees without cones, leaving more room for ornaments.

All of these tasks must be performed manually, tree by tree, tens of thousands of times per year—an enormous but largely unseen effort behind this romanticized holiday tradition. By the time a tree reaches harvest, workers have typically visited or passed it or touched it with their hands more than 150 times.

Successful harvest timing hinges critically on cold weather, which ensures that needles adhere firmly to branches. Trees cut during unseasonably warm weather risk losing their needles and becoming unmarketable. Christmas tree farmers like Stone must therefore wait for precisely the right seasonal cold snap to begin the frenetic harvest each year.

Two days before harvest, Tree Kings crews move methodically through fields, armed with multicolored sizing poles, evaluating every tree and tagging those ready for market. Immediately behind the taggers and cutters, farmworkers sling harvested trees onto nine-hundred-pound-capacity rope hammocks. Moments later, contracted helicopters roar overhead, sending powerful gusts whipping through the fields as pilots deftly maneuver to secure their loads. Tree hammocks sway precariously beneath choppers as they are airlifted toward the central processing area. There, trees get unloaded and sent one by one through a massive vibrating machine, which shakes the trees for fifteen seconds to dislodge dead needles. Then they go straight into a spinning baler, which wraps branches in protective mesh and twine. Trucks depart around the clock, ferrying thousands of trees to merchants nationwide. During the blistering thirty-day sprint leading up to Christmas, large operations like Tree Kings sell upwards of two hundred thousand trees to big-box retailers, whose discount-driven strategies have steadily squeezed smaller growers to the economic margins.

Every holiday season, Stone walks a tightrope, aiming not to harvest too many trees, lest he jeopardize future profitability by removing younger trees before their maximum market value is realized. That may seem straightforward in theory, but it is more difficult when a sizable purchase order comes in. The temptation to cut just one more tree can be hard to resist but could be financially ruinous down the line.

Despite the magnitude and spectacle of the white-knuckled, labor-intensive Christmas tree hustle, the economics remain precarious at best. Half of American Christmas tree farms report gross revenues under $25,000 annually, making the endeavor essentially a part-time hobby for most small growers, who hold off-farm jobs for supplemental income. Even at large farms like Tree Kings, profit margins rarely top 9 percent. Nitrogen fertilizer prices have more than doubled since Russia's invasion of Ukraine, and machinery costs are similarly exorbitant. A large farm might need a dozen tractors, each costing around $80,000 and running on $4 per gallon diesel. Add in the $70,000 baler machine, and production costs escalate. In a typical year, between 5 and 10 percent of seedlings planted will die and must be replaced with fresh seedlings purchased on nursery contracts with three-year lead times.

All told, Tree Kings' production cost for a single six-foot tree runs around $16. Stone wholesales these trees to retailers like Home Depot and Lowe's for about $25 each, which then turn around and sell the same tree to consumers for $75. Ultimately, the farmer responsible for all the hard work earns half the profit margin enjoyed by retail giants with gross annual revenues exceeding $150 billion.

"Home Depot sets the price, and the rest of the market follows. I only get paid for what they actually sell," Stone told me, shaking his head. "Early on in my career, we'd start shipping trees the first week in December and work right up through Christmas. Now I make my final shipment December fifth and then fly around to Home Depots in Vegas and LA and Dallas, pushing trees."

The daunting economic reality of Christmas tree farming reflects a broader shift in American culture. As natural resources became increasingly scarce and costly during the late twentieth century, consumer tastes began to gravitate toward engineered, synthetic, and artificial alternatives. Global competition among consumer-products corporations triggered an unprecedented wave of technological innovation and creative marketing that turned scarcity into opportunity, manipulating America's relationship to authenticity and nature itself.

Faced with soaring lumber prices and limited supplies of preferred species like western red cedar and Douglas-fir, lumberyards turned toward cheaper, more abundant evergreens, obscuring their inferiority with vague branding. Stacks labeled HEM-FIR often meant mixtures of structurally weaker species like western hemlock and balsam fir. Another economical choice was marketed simply as SPF (spruce-pine-fir), combining various conifers harvested indiscriminately and sold interchangeably. Perhaps most absurdly, lumberyards introduced "white wood," an ambiguous category covering low-grade lumber typically composed of Baltic, Scots, or Norway pines. Whether or not it was intended, white wood echoed other twentieth-century branding triumphs: white bread, white wine, and white meat. Precise distinctions between evergreen conifer species that botanists had meticulously defined over centuries were brushed aside to suit the homogenizing, mass-produced American marketplace.

The newfound technological efficiency at sawmills continued producing enormous amounts of waste in the form of sawdust, small-diameter trunks, and unsalable scraps. But rather than discarding or burning these byproducts as generations had done before, inventors sought creative ways to transform them into marketable goods. In 1925, William Mason, a friend of Thomas Edison, set out to find a use for mountains of wood shavings from lumber mills. One evening, Mason accidentally left his equipment running overnight, and by morning he had discovered a thin yet surprisingly durable sheet of compressed fibers. He quickly patented his accidental innovation under the brand name Masonite. Manufacturers refined Mason's method, combining pulverized evergreen fibers bonded by urea-formaldehyde resins under extreme heat and pressure to create medium-density fiberboard (MDF). By the 1980s, MDF had become a household staple, commonly used for kitchen cabinets and inexpensive furniture, which were paradoxically cheaper than the same products made from traditional lumber or plywood. Though MDF production was energy-intensive, relying heavily on chemical resins and fossil fuels—and drawing scrutiny from the Environmental Protection Agency for formaldehyde emissions—its rise highlighted an environmental paradox.

Engineered wood products became ubiquitous precisely because genuine evergreens had grown scarce. Viewed against this new reality, the simple saltbox homes of colonial America seemed luxurious—a recognition perhaps overdue.

As the postmodern American economy evolved alongside shifting consumer tastes, traditional building materials were increasingly supplanted by synthetic alternatives that attempted to outperform Mother Nature. Cedar shingles and clapboard siding faced stiff competition from artificial facsimiles made of vinyl and asphalt. Interior plywood sheathing yielded to gypsum-based drywall marketed as Sheetrock. Knotty pine floors were smeared with glues and covered over by shag carpeting, ceramic tiles, and linoleum. America's growing preference for engineered and artificial materials was not limited to construction but extended into every aspect of domestic life, reflecting a cultural shift toward convenience, uniformity, and manufactured perfection.

At the same time, cheap evergreen pulp fueled an explosion of consumer products once inconceivable, yet now seemingly indispensable—disposable plates, coffee cups, and diapers all emblematic of a throwaway culture dependent upon abundant yet invisible forest resources. Eggs and milk started showing up in cardboard packaging. Kotex introduced the feminine sanitary napkin, and Kleenex launched its eponymous facial tissues. Everything from breakfast cereal to motor oil came packed inside at least one cardboard box from the moment it left the factory to its arrival on American doorsteps. Paper also supported an explosion in record-keeping, business documentation, government bureaucracy, publishing, and mass media, facilitating a freer flow of information to the body politic.

Our reliance on paper extended into the refinement of cellulose acetate, a thermoplastic derived from lignocellulose, extracted from the pulp of evergreen conifers. Cellulose acetate fibers were sewn into shimmering prom dresses and synthetic textiles like rayon and spandex, and enabled manufacturers to produce everyday consumer items like eyeglass frames, ballpoint pens, photographic film, cigarette filters, playing cards, pantyhose, Lego bricks, imitation leather, and liquid-crystal display television screens. Food manufacturers also embraced synthetics, adding powdered wood flour as an undigestible thickening agent in yogurt and ice cream. As American consumerism evolved, the presence of evergreen materials grew pervasive yet invisible: a hidden abundance woven seamlessly into everyday life.

Evergreens were everywhere, yet hardly anyone noticed.

A collective fixation on artificiality and engineered perfection ultimately reached the Christmas tree itself. Long before contemporary plastic evergreens dominated the market, artificial Christmas trees underwent decades of awkward experimentation and misguided ingenuity. Early attempts date to 1913, when Sears, Roebuck and Co. sold its first fake tree: a spare assemblage of dyed goose feathers attached to wire branches. In the 1930s, Addis Housewares, famous for its badger-hair toilet brushes, ingeniously attached green-dyed toilet brushes together, creating prickly, evergreen-shaped decorations. Thousands of these bristle-brush trees were shipped to England to lift spirits after Nazi bombing raids devastated London.

Other inventors pursued riskier ideas, like coating live evergreens with aluminum paint using an insect-spray gun. This practice quickly proved hazardous, combining dry pine needles, oil-based paint, indoor smoking, and gas heating into a ticking time bomb for American living rooms. Recognizing the dangers, the Aluminum Company of America introduced a safer alternative in 1959. Dubbed the Evergleam, this sleek metallic tree featured foil needles and sold for twenty-five bucks. Alcoa immediately sold out of its initial run of 10,000 trees and devoted entire production lines to meeting demand. By 1964, Evergleam was selling 150,000 aluminum trees annually, complete with rotating color disks projecting lights reminiscent of *Saturday Night Fever* disco balls. Yet the Evergleam's dazzling appeal carried a serious hidden danger: the ever-present risk of electrocution. The recently formed US Consumer Product Safety Commission issued this blunt warning in 1978: "Never use Christmas lights on a metallic tree because a person could be electrocuted if the tree becomes charged with electricity . . ." Safety concerns, coupled with shifting consumer tastes, eventually doomed aluminum trees, and Alcoa ceased Evergleam production entirely. Some were rescued from attics as vintage collectibles; most, however, were unceremoniously discarded in landfills.

Few people noticed when, in 1972, a former World War II bomber pilot named Si Spiegel patented an innovative Christmas tree designed with lifelike plastic needles and hinged branches for easy assembly. Spiegel saw untapped potential: Bristle-brush trees looked obviously fake, Evergleams posed safety risks, and real trees shed needles, oozed sap, and dried into fire hazards. His solution? A tree that never dies.

Spiegel's boss at American Brush Machinery in Mount Vernon, New York, had allowed him to repurpose its brush-making machines to create artificial Christmas trees using green polyvinyl chloride (PVC) plastic. The first attempts hardly resembled real evergreens, and sales were sluggish. Few mid-century households yet understood the appeal of plastic replicas. Sent to shut down the failing factory, Spiegel instead returned convinced the company could profit by refining its artificial trees. One boss dismissed Spiegel as crazy, but another believed in him enough to grant him control of his own division: American Tree and Wreath.

Determined to perfect the product, Spiegel studied real evergreens and adjusted the machinery to streamline production. His redesigned trees were remarkably realistic, affordable, and quick to manufacture. By the mid-1970s, American Tree and Wreath produced eight hundred thousand trees annually, with one fake tree rolling off the assembly line every four minutes.

Time magazine caught wind of the trend early, running the headline AND A PROFIT IN A POLYVINYL TREE in 1964, noting that fake trees looked more realistic than anyone imagined possible. Yet these artificial trees originated neither from snowy tree farms nor serene evergreen forests, but from oil rigs drilling beneath the ocean floor, extracting petroleum to be refined into plastics at factories half a world away.

American enthusiasm for artificial Christmas trees surged after scientists at DuPont discovered that polymers derived from not just PVC but also petroleum-based polyethylene terephthalate (PET) could be shaped into consumer staples to fulfill almost every human desire. These versatile plastics painted a utopian vision of a future abundant in material wealth: safe, inexpensive, and endlessly moldable. Plastics replaced steel in cars, paper in packaging, wood in furniture, and eventually the living conifers in American homes.

Plastic trees evolved over the ensuing decades into hyperrealistic replicas, featuring prestrung LED lights, Bluetooth speakers, revolving bases, and even a spritz of synthetic pine scent. These newfangled imposters were marketed as environmentally virtuous objects sparing natural forests from seasonal destruction. By 2023, *The New York Times*'s Wirecutter guide to artificial Christmas trees had become its most-read product review, out of a catalog of more than one thousand consumer items.

Most artificial Christmas trees originate not in Santa's workshop at the North Pole but in the true central hub of America's holiday industrial complex: Yiwu, China. This sprawling industrial city—devoid of snow, elves, or genuine evergreens—is home to a four-mile-long stretch where six hundred factories churn out 60 percent of the world's Christmas decorations. Here, migrant laborers, the factory "elves," work twelve-hour days dipping polystyrene "pine" branches in bathtubs of glue and spraying fake snow on them by the millions. Christmas is an abstract idea to these workers; one remarked to the Chinese state news agency, "Maybe it's like New Year for foreigners." *The Atlantic* has reported that Santa Claus is more familiar than Jesus Christ to most people in China.

A labyrinthine complex of sixty-two thousand storefronts in Yiwu shows off this landfill-choking wonderland of plastic Christmas paraphernalia. Whole neighborhoods throb with plastic pine cones flocked in gold and silver, and floor-to-ceiling merchant booths feature plastic Christmas trees in yellow and red and glittery pink. Wholesale buyers from Europe and the Americas convene in this retail megacity every February to get a sneak peek of the season's hottest trends, ensuring that artificial trees and decorations reach North American ports by June, and can stock warehouses by early October. The plastic Christmas tree supply chain offers corporations stable and predictable inventory management, an advantage decidedly absent from the compressed season faced by fresh-cut evergreen farms.

Perhaps no one has capitalized more on the plastic tree supply chain than Mac Harman, a former McKinsey consultant who turned the Christmas tree industry upside down when he founded Balsam Brands in 2006. Harman has said he got the idea to sell artificial trees after visiting his in-laws over the holidays and recoiling at their "sad, cheapo" fake tree. In his early market research, he learned that 90 percent of the $1.2 billion artificial Christmas tree market was captured by mass retailers like Walmart and Home Depot, and he saw room for a more upscale product sold directly to consumers.

Harman was the first to recognize that artificial Christmas trees had begun looking unnaturally pristine and symmetrical. To rekindle authenticity and emotional resonance, he collaborated with Chinese manufacturers using live evergreens as templates, deliberately incorporating imperfections like crooked branches and color variations that closely mimicked real trees' natural quirks. Harman's innovative approach struck gold by making fake trees appear convincingly less fake. One of Balsam Brands' most popular models—a seven-and-a-half-foot tall noble fir with LED lights priced at $1,249—captivated consumers and took the holiday market by storm, challenging real-tree farmers like Tyler Stone at every turn.

Whether intentionally or not, Harman's visionary product cleverly exploited Americans' conflicted relationship with authenticity. By selling mass-produced, realistically flawed artificial trees, Balsam Brands effectively rendered real evergreens a quaint inconvenience—an impractical luxury. In this contemporary narrative, the blamelessness of plastic was affirmed while real evergreen conifers acquired an inexplicable moral stain that bankrupted thousands of family farms and injected plastic into an untold number of perfectly good Christmases.

Centuries ago, William Shakespeare cautioned against unnecessary embellishments in the fourth act of *King John*, when the Earl of Salisbury warned against a second coronation, saying:

> *To gild refined gold, to paint the lily,*
> *To throw a perfume on the violet,*
> *To smooth the ice, or add another hue*
> *Unto the rainbow, or with taper-light . . .*
> *Is wasteful and ridiculous excess.*

Today's artificial-Christmas-tree industry excels precisely in this sort of gilded embellishment, spraying glitter and synthetic pine fragrance onto plastic branches and selling the elaborate fakery as "authentic." Real evergreens—ancient symbols of longevity and fertility—have been portrayed as an environmentally irresponsible choice, even as local tree farmers quietly sustain traditional agriculture in a society that does not value or understand its importance.

Despite their commercial success and the veneer of convenience, plastic trees come with stark environmental costs. Most of them end up discarded after ten years, joining the three hundred million tons of plastic waste generated globally each year, an amount not far off from the collective weight of the entire human population. Satellite imagery has captured vast floating plastic debris fields, including Christmas decorations, drifting through the Pacific Ocean, where the Great Pacific Garbage Patch churns over an area twice the size of Texas. On remote Midway Atoll, thousands of miles from major population centers, wildlife biologists have found seabirds nesting amid discarded artificial tree branches and glittery plastic ornaments, grim reminders of our distant celebrations.

Most plastic waste breaks down into tiny shreds called microplastics, which scientists have detected in places as remote and inaccessible as the deepest ocean trenches and the summit of Mount Everest. Microplastics are uniquely insidious due to their dual role as carriers and releasers of toxic chemicals. Derived from petroleum, plastics contain known carcinogens, such as benzene and vinyl chloride. As they degrade into microplastics, these toxins leach into water, soil, and air. Simultaneously, microplastics act as magnets for persistent environmental poisons, including pesticides and industrial chemicals. Marine life consumes these particles, which ultimately enter human bodies through breathing air, drinking water, and eating fish.

Research now clearly shows microplastics accumulating in human blood, breast milk, semen, and brain tissue. Infants are exposed before birth; microplastic particles have been measured in startling concentrations in human placentas and newborns' first stool samples. A 2024 study in a leading toxicology journal suggested that microplastic exposure may explain the alarming global decline in male reproductive fertility observed in the twenty-first century.

Once celebrated for their environmental virtue, plastic trees now represent an ironic legacy: synthetic imitations polluting the natural world they originally sought to protect. Remember, plastic was beloved precisely because it stuck around forever. Plastic is evergreen too.

Like many agricultural decisions, the choice each Christmas boils down to two essential options: support local farmers who nurture living evergreens and vibrant communities, or opt for distant industrial manufacturers whose methods burden the planet with pollution and compromise human health. When this choice becomes overwhelming amid the cacophony of glitter and twinkling lights, authentic desires fade, preferences weaken, and consumers struggle to answer a fundamental human question: *What do I really want?*

In this existential debate of real versus fake, pine versus plastic, natural versus synthetic, the authentic Christmas tree, grown by human hands and harvested from living soil, remains profoundly meaningful.

Bringing a living evergreen into our homes connects modern Americans directly to something primal and essential: our shared and complicated history. Real trees put us in touch with all that we share with other creatures and all that sets us apart. They show us a mirror.

What is perhaps most troubling, and sad, about the modern plastics industry is how effectively it obscures the profound connections and relationships embedded in our holiday traditions. To go from the prehistoric ancestor of evergreens, Archaeopteris, to a glitter-flocked, Bluetooth-equipped, perfume-spritzed plastic tree is to leave this world on a journey of forgetting where we came from. But forgetting, or not knowing in the first place, is exactly what the holiday industrial complex expects of us. For if we could see the truth about our festive customs, we would almost certainly change the way we celebrate.

Though much has been done to obscure this simple fact, how we shop largely determines our role in this world and what will become of it. To shop with a full awareness of all that is at stake for humanity might seem burdensome, but few things in life could be more satisfying. Many people today seem perfectly content shopping at the end of a plastic supply chain without a thought in the world. Real evergreens are probably not for them. We know things about the trees' past that could spoil the holiday cheer. But in the end, this is a book about the many wonders of Christmas trees, the kinds of wonders that are only deepened by knowing.

Every December, millions of trees groomed on farms like Tyler Stone's take center stage in homes, offices, and town squares around the world. Weeks later, as needles drop and celebrations fade, each tree reaches the natural conclusion of its journey. Whether chopped into fragrant mulch for a community garden, burned in a crackling winter bonfire, submerged in a lake as habitat for spawning fish, or simply set curbside on a frosty January morning awaiting collection, the evergreen fulfills its fundamental purpose. In this final transformation, it performs the most evergreen act of all.

It returns to the earth.

22

All from a Seed This Big

▲ ▲ ▲

CATASTROPHES HAVE ALWAYS happened, and always will. People long for solutions to the grand challenges of their era, though no single generation can resolve them completely. Sometimes the remedies proposed become heavier burdens than the problems they seek to solve. Replanting deforested landscapes may never fully restore the richness that has been lost, but even a handful of dedicated people carrying buckets of seedlings can help reclaim areas decimated by centuries of human folly.

Still, reforestation is easier said than done. To appreciate the challenge of cultivating resilient seedlings, consider the curious sex life of evergreen conifers. Most species produce no seeds until their fifteenth year, and viable seed crops typically form only once every seven years thereafter. What they lack in frequency, they compensate for in abundance, with the average conifer producing roughly ten million seeds over its lifetime. The seeds form after male cones release pollen, which is carried by wind until it lands on female cones, where fertilization takes place. The Monterey pine (*Pinus radiata*) holds the record for prolific pollen release, with each tree discharging up to two pounds annually—a messy botanical ejaculation responsible for countless allergic reactions every spring. California's Coulter pine (*Pinus coulteri*) produces female cones weighing up to eleven pounds each, which drop with the force of a bowling ball, earning the tree its nickname: widow maker.

Male cones, clustered inconspicuously on the lower branches, are small, soft, and short-lived, while sturdier female cones typically adorn the upper canopy. The clever spatial separation minimizes self-pollination, since gravity rarely carries pollen upward on the same tree. Some species further reduce the risk of self-fertilization by staggering the maturation times of male and female cones, thus preserving genetic diversity.

After fertilization, female cones remain sealed and rock-hard for up to three years, their scales bound by resin as strong as superglue, protecting the developing seeds within. About 88 percent of conifer species exhibit some degree of serotiny, a remarkable evolutionary adaptation whereby mature female cones remain sealed for decades until a wildfire heats them above 122°F. At this critical temperature, the resin melts like candlewax, cracking open the scales to release delicate, paper-winged seeds. Some float away with a gentle hiss, carried by the breeze, while others spin down into the nutrient-rich, fire-cleared soil. These seeds germinate and grow fast in their youth, giving evergreens a crucial evolutionary advantage over competing plants. In the absence of fire, serotinous cones can still open and release their seeds, but only after natural decomposition gradually weakens the resin over the course of several years.

Seedling establishment forms the bedrock of ecological succession, the centuries-long rebuilding of complex ecosystems after natural or human disturbances. On the East Coast, for instance, deforested lands untouched by severe wildfires have made remarkable recoveries. Roughly 80 percent of coastal New England is again blanketed by trees, in what amounts to a massive undoing of the many excesses of the industrial age. As farmland was abandoned and younger generations left rural communities, fallow fields gradually reverted to forests, reclaiming landscapes cleared centuries earlier by colonial agriculture and industry. Although these regenerated second-growth woodlands remain fragmented by highways and suburbs, and compromised by invasive species, they represent a hopeful example of nature's resilience when given the chance to heal.

In the western US, however, the natural processes that drive seedling regrowth have been severely disrupted by decades of intensive logging, fire suppression, and climate change. A 2019 study in the *Proceedings of the National Academy of Sciences* warned that many western forests have already crossed ecological tipping points, beyond which natural recovery becomes impossible. An extensive survey of thirty-three wildfires revealed that ponderosa pine and Douglas-fir seedlings struggle mightily to establish themselves in heavily burned or logged areas. Without healthy seed-bearing trees near burn sites, these forests may permanently lose their capacity for regeneration.

Such existential threats underscore the urgency of human-led reforestation. Seed banks and plant nurseries have become critical lifelines, stepping in where natural regeneration falters. At California's Lewis A. Moran Reforestation Center in Davis—operated by the California Department of Forestry and Fire Protection—the future of western forests is quietly taking root. Founded in 1921 in response to widespread logging and wildfire, the Moran Center is among America's oldest continuously operating forest nurseries. Here, the California Cone Corps is diligently untangling a century of forest mismanagement to restore landscapes that human blunders nearly destroyed. The group's efforts represent more than ecological renewal; they are a reckoning with history.

For over a century, Cal Fire's guiding principle has been to plant seeds from "fire-hardened survivors, which possess the genes we want to preserve and reproduce in localized seedling populations," the Moran Center's seed bank manager, Denia Troxell, told me. These carefully chosen "plus trees" carry superior genetic traits adapted to resist diseases such as pine blister rust and infestations of devastating bark beetles. Within Cal Fire's valuable cache of climate-resistant plus trees, ten high-priority conifer species have been identified, their precise geolocations noted for ongoing seed collection.

"Small family forest owners can't get seedlings to replant their burned-over lands from large timber companies," explained Britta

Dyer, the energetic vice president for resilient forests at the conservation nonprofit American Forests. "You'd need to order ten thousand trees and grow them on contract. That model is radically different from public lands. Our mission is to reawaken the Forest Service to these methods, teaching agencies how to rebuild a seedling supply chain with new science and renewed care."

Despite their immense economic importance, seedling nurseries remain largely invisible to the public, and few people outside forestry circles have set foot in one. Some of this obscurity stems from the industry's wariness of outsiders, some from widespread indifference to the true origin and cost of wood products. Mostly, though, the painstaking work unfolds quietly in remote greenhouses far from urban life.

At the Moran Center, decades-old equipment and neglected facilities reflect chronic underfunding, yet a passionate staff continues this vital work, sustained by a modest annual budget of just $1.2 million. The morning I visited in June 2023, crews harvested fifty bushels of seed cones, a haul that seemed substantial until Natalie Kim, a Cone Corps nursery aide, offered a sobering perspective. "To plant just 25 percent of privately owned land needing reforestation, we must collect over seventy-six thousand bushels of cones," she said. That day's harvest amounted to 0.01 percent of their annual goal, since federal guidelines require storing enough seeds for each species to last a decade.

Collecting cones sounds simple—pluck a cone, drop it in a bucket—but the reality can be life-threatening. The freshest, seed-rich cones grow at treetop heights, where the slightest breeze exaggerates branch movement, whipping limbs like fishing rods mid-cast. In the freewheeling 1960s, cone collectors blasted upper branches off trees with shotguns. Today, eclectic teams of rock climbers, skydivers, window washers, and thrill seekers scale dizzying heights, each equipped with personal liability insurance to harvest cones from wild forests.

Modern seed orchards now offer a safer, more efficient alternative by pruning elite trees to lower heights for easier cone picking. Forestry scientists have also developed methods to graft young shoots onto mature rootstock or notch trunks to mimic insect damage, stimulating reproductive hormones—a botanical sleight of hand that accelerates nature's timeline.

After cones are either plucked precariously from treetops or gathered methodically from pruned seed orchards, they undergo a time-honored extraction process. Workers empty burlap sacks of cones onto wooden trays, then stack them inside rustic kilns whose walls bear the blackened scars of countless drying cycles. Technicians carefully regulate temperature and humidity, circulating warm air until the resin melts and the cone scales crack open, revealing their precious cargo. Rattling contraptions reminiscent of antique wheat threshers then tumble the cones, shaking loose their seeds, which cascade through multiple screens to separate seeds from chaff. Machines known as the scalper and the dewinger remove residual debris before the purified seed batch undergoes x-ray screening to confirm embryo viability.

Everything about this bootstrapped operation evokes the nineteenth century: belts, rotors, and lubricated wheels. Nursery technology has remained essentially frozen in time compared to modern agriculture, constrained by practical limitations and budgets. The equipment could be two centuries old, yet it remains state-of-the-art. Even today, the seed batches are tracked with beautifully simple tools: paper labels and Sharpie pens.

Processed seeds soak in water for two days before entering cold-storage lockers kept at 6°F, mimicking the dormant chill of winter. Stepping inside, breath crystallizes, and ice faintly crackles around door hinges. The room resonates with scents of vanilla and turpentine, sharp enough to recall piney cough drops: fresh, intense, sublime. Floor-to-ceiling shelves hold thousands of sealed plastic bags, each one containing a future forest, patiently awaiting the right moment for its DNA to awaken and thrive.

From the freezer, seeds move into greenhouses, where technicians plant them individually in slender pots shaped like Cuban cigars. Delicate seedlings less than a foot tall stand in neat rows, fragile yet filled with potential. In nature, a single five-hundred-year-old evergreen might yield only a handful of seedlings that survive to this stage. Here, their odds of survival dramatically improve, though Mother Nature makes no promises.

There are no shortcuts or hacks for tree planting. It cannot be ordered casually from a smartphone nor summoned by artificial intelligence. Reforestation still requires real people willing to scale remote mountainsides under harsh conditions. Recruiting for these roles gets harder every year, but those best suited often come from Native American communities, whose connection to the land runs deep.

From California, the seedlings' journey continues north and east to Montana's Flathead Indian Reservation, where a restoration project exemplifies ecological renewal intertwined with cultural revival. Many tribes across North America have resurrected traditional practices like controlled burning to heal damaged forests. The Confederated Salish and Kootenai Tribes, however, have prioritized greenhouse cultivation to support conifers struggling to regenerate naturally. Among these is the slow-growing whitebark pine (*Pinus albicaulis*), its population severely threatened by invasive blister rust, aggressive bark beetles, and intensified wildfires. With the longest seed maturation cycle among North American conifers—sixty to eighty years—replanting whitebark pines protects more than a species. It safeguards an entire worldview.

"Everything we plant now is for our grandkids," ShiNaasha Pete, the tribe's reforestation forester, has said, voicing a sentiment echoed by Indigenous communities nationwide. "Whitebark pine was a supplement to the original diet of our ancestors. We want to keep our culture and our language and lifestyle alive in our younger generations."

After identifying a handful of plus trees genetically resistant to blister rust, Pete's team collected enough seeds to eventually repopulate the tribe's entire 105,000 acres of whitebark pine habitat. They began in 2023 by planting more than four thousand seedlings, aiming to scale up to fifty thousand per year. Even as many states cut back or shutter seedling nursery programs, two other Montana tribes—the Assiniboine and the Gros Ventre, of the Fort Belknap Indian Community—have launched projects to restore forests of subalpine fir (*Abies lasiocarpa*) wiped out by wildfires in the 1930s, using federal infrastructure funds to bring the trees home again.

Across the United States, tribal nations now oversee more than fifty-six million acres of reservation lands reclaimed after centuries of dispossession. Many have formed partnerships with government agencies and conservation groups to replant fire scars within their territories, while others grow seedlings for broader regional efforts.

In New Mexico, the ambitious John T. Harrington Forestry Research Center at Mora collaborates closely with Indigenous communities to incorporate traditional knowledge into modern reforestation methods. Its goal is bold: producing five million seedlings annually to revitalize fire-ravaged southwestern forests through planting initiatives led by Native Americans. The Jemez Pueblo and the Santa Clara Pueblo in north central New Mexico have joined forces with The Nature Conservancy to plant over one hundred thousand drought-resistant ponderosa pine seedlings in the Jemez Mountains, where the Las Conchas Fire of 2011 left a vast ecological scar. New Mexico's Mescalero Apache Tribe operates another vital restoration program focused on ponderosa pine and Douglas-fir—two trees imbued with profound cultural significance. During Spanish colonial rule, the Mescalero people were enslaved and forced to cut down these very species to build Catholic missions throughout the region. Today, they grow about seventy thousand seedlings annually to plant in the Lincoln National Forest. For the Mescalero Apache, reforestation is an act of remembrance.

Yet these substantial efforts represent less than 1 percent of what is truly needed. New Mexico alone will require at least 150 million seedlings to fully restore forests lost to wildfires. The sheer scale of this challenge underscores both the urgency and the complexity of the path ahead.

Those spearheading reforestation efforts across the West descend directly from Indigenous peoples who inhabited America's grand forests long before European colonization fractured their livelihoods and cultures. Their approach to ecological restoration surpasses conventional forestry, blending scientific rigor with ancient wisdom. Every seedling nurtured and every sapling rooted in ancestral soil reclaims a piece of their heritage while renewing the landscape. Through these quiet acts of stewardship, tribal nations ensure that the land and its sacred sites will flourish once again for future generations. Perhaps this represents the most enduring truth of all: Hope lives within the tiny, evergreen promise of a seed.

EPILOGUE

Have Yourself a Garish Little Christmas

THE PARKING LOT at my local Christmas tree farm was already overflowing when I arrived. A stream of shoppers moved toward the entrance, the air humming with snippets of Mariah Carey's "All I Want for Christmas Is You" from distant speakers. It was mid-December, and the holiday frenzy had reached its tinsel-laden peak. I hesitated in the car, hands gripping the wheel, unsure if coming here had been a mistake.

After weeks of deliberately avoiding Christmas—the decorations, the music, the whole exhausting spectacle—I now stood squarely in the heart of America's holiday industrial complex, a place I had sworn not to go. But after nearly a year stuck indoors, dulled by pandemic boredom, the pull of tradition in its gaudiest form proved irresistible.

I got out and zigzagged through the parking lot, past screaming children and their sleep-deprived parents, and stepped through the straw-bale gates into Santa's psychedelic workshop: part kaleidoscopic amusement park, part state fair, part garden center, part fever dream. Teenagers circled an ice-skating rink, wobbling beneath oversized Santa hats. A food truck selling peppermint hot chocolate pumped sugary steam into the chilly air. Off in one corner, reindeer placidly chewed hay, seemingly unbothered by the commotion.

And, of course, there were trees. Thousands of them. Rows upon rows of spruces and firs, ranging from twenty-foot Leviathans to one-foot tabletop centerpieces. Baskets, boughs, ribbons, ornaments, pine cones, and wreaths crowded every aisle. This place had everything and more.

I turned my attention to choosing a tree but quickly realized these were not ordinary evergreens. Here stood spruces and firs painted candy-apple red, shimmering metallic gold, electric bubblegum pink, and icy blue, a palette worthy of a Lisa Frank binder or Elsa's dresses from *Frozen*. They could not possibly be real. I stepped closer and pinched a branch, expecting it to feel plastic. But no. The needles were pliant, soft, and alive—real trees cloaked beneath a festive fantasy. Their natural green had been erased by our impulse to make nature more palatable, more marketable, more *artificial*. Nothing says Christmas quite like a Dr. Seuss conifer that looks freshly plugged into a light socket.

"We Americans like fake things, right?" asked a young employee dressed as the Grinch, catching my incredulous stare. He made a cupping gesture with both hands. "Boobs, Christmas trees, you name it—everybody wants to be special."

He told me the color trend had taken off in 2010 at the Kirk Company, a wholesaler in Wautoma, Wisconsin, the self-declared Christmas tree capital of the world. Initially, the company produced green, fireproof latex paint to help farmers hide unsightly brown patches on otherwise beautiful trees. Then someone added pigments, and the trend spread like wildfire. The only drawback, the employee admitted, was that the paint dulled the natural piney scent—ironically, one of the main reasons people wanted a real tree in the first place.

I lingered beside a cluster of brilliant-teal Douglas-firs, shaking my head. All this, I thought, in the name of Christmas. Planes fly in from China. Trucks roll out of Oregon and Wisconsin and North Carolina. Pop-up shops sprawl across acres of parking lots. Billions of dollars change hands, all for the wreath casually tossed in a Trader Joe's cart, for the glass Santa ornament mailed to cheer up someone's mother, for the tree that will sit forgotten at the curb on January second.

Overwhelmed by the spectacle, I retreated to a quieter corner of the nursery, passing metal carts of half-price wreaths and a booth overflowing with inflatable snowmen. Tucked discreetly in the back, I finally found what I was looking for: the nursery's modest collection of potted living evergreens. They stood apart from the glittering chaos, quiet and dignified. I spotted a solitary Colorado blue spruce (*Picea pungens*), its needles tinged a silvery frosted blue, gracefully curving upward. This was the one.

When I got home, I lugged the seven-gallon pot from my trunk, wheeled it around the side of the house, and set it outside the backyard window. I mean, who really wants dry needles scattered across the living room floor? Planting it was not easy. The shovel kept jamming against rocky soil, and my shirt grew damp with sweat. The unseasonably warm weather felt like a quiet rebuke, a subtle reminder of the stakes of our modern excesses.

After wrestling the spruce into its permanent home and tamping down the soil, I stepped back to inspect it from every angle. It did not need much. A single strand of lights gave it just enough glow. Standing there, still holding the shovel, I realized how profoundly different this felt from any tree I had ever brought home. No extravagant colors. No plastic imitations. No crooked balsam firs cut and dying in the corner. Just something simple, maybe even everlasting—a tree that will probably still be sending roots deep into the earth and stretching its limbs skyward long after I am gone. Maybe a hundred Christmases from now, someone will gaze up at it and marvel that the world can still hold something so big and timeless.

I reached out and pinched its needles between my fingers, breathing deeply, savoring that unmistakable piney scent. It carried faint traces of wildfires and sawmills, lumberjack bunkhouses and bustling home construction sites. Hints emerged of turpentine and damp wool trousers, ancient temples and coastal bogs, battlefields and steamboats, iron furnaces, railroads, and sailing ships. The fragrance evoked Christmas tree lots and wartime soldier barracks, decomposing slash piles and distant forests, and all the lives evergreens had ever touched, breathing the same crisp air across generations.

Every tree told a story.

Every forest held history.

Even after everything I had seen, I was still swept away by the magic of it all.

ACKNOWLEDGMENTS

I SPENT DAYS paralyzed by how many people I wanted to thank before deciding to just thank everyone, because I did not want to leave anyone out. The last time I wrote a book, I offered a laundry list of wonderful people spread over pages and pages, and *I still* forgot some important names. You guys know who you are, and if I have been any kind of decent friend or colleague, you ought to know why I appreciate your presence in my life. But I am especially grateful to Adam Chromy, my agent at Movable Type Management, Nadxieli Nieto, editorial director at Algonquin Books, and Sally Kim, president and publisher at Little, Brown and Company, without whom this book would not exist. Agents and editors—including the ever-helpful Jovanna Brinck, Elizabeth Johnson, and Brunson Hoole—always seem to have compelling ways of shaping my tangled notes and reflections into a coherent narrative.

BIBLIOGRAPHY

WHILE WRITING THIS book, I avoided obstructing the reader's experience with exhaustive bibliographical citations, but I remained meticulous about my sources. The quotations and facts presented here are drawn from the works listed below, alongside a rich variety of published literature. Equally essential were insights gathered through personal interviews, conversations with experts, and my own direct experiences in forests and fields across America.

My personal library has been both refuge and revelation, and numerous discussions with scientists, historians, and thoughtful friends have deepened my understanding. During my travels and research, I encountered ecologists, foresters, Indigenous knowledge holders, tree farmers, and other keen observers of the natural world, whose generosity and wisdom proved invaluable.

The accelerating pace of environmental change made keeping this narrative current a constant challenge. To stay abreast, I relied heavily on authoritative and up-to-date sources such as *Science* and *Nature*, along with indispensable reporting from outlets like *The Guardian*, *The New York Times*, *The Washington Post*, and *The Economist*, and excellent journalism from many others.

My hope is that readers will gain a deeper appreciation for how evergreens underpin so many aspects of American life, while perhaps also finding some comfort in knowing that catastrophic upheaval has always been our planet's way of doing things.

Alagona, Peter S. *The Accidental Ecosystem: People and Wildlife in American Cities.* University of California Press, 2022.

Apps, Jerry. *When the White Pine Was King: A History of Lumberjacks, Log Drives, and Sawdust Cities in Wisconsin.* Wisconsin Historical Society Press, 2020.

Arno, Stephen F., and Carl E. Fiedler. *Douglas Fir: The Story of the West's Most Remarkable Tree.* Mountaineers Books, 2020.

Baldwin, William C., Angela Chiang, Gabrielle Wyatt, et al. "Distinct Terpene Metabolite Blends Serve as Core Drivers of Consumer Aroma Preferences in Fir Christmas Tree Species." *Scientific Reports* 15 (2025): 18500. https://doi.org/10.1038/s41598-025-01505-1.

Barham, Lawrence, Geoffrey A. T. Duller, Ian Candy, et al. "Evidence for the Earliest Structural Use of Wood at Least 476,000 Years Ago." *Nature* 622 (2023): 107–11. https://doi.org/10.1038/s41586-023-06557-9.

Bartram, William. *Travels Through North and South Carolina, Georgia, East and West Florida, the Cherokee Country, the Extensive Territories of the Muscogulges, or Creek Confederacy, and the Country of the Chactaws.* James & Johnson, 1791.

Bettis, W. M. Letter to Olin Colip, 1951. In Kathleen A. Brown's reply to Seth Wigderson's query in the discussion titled "Homosexuality in Logging, Mining, and Railroad Camps." H-Net: Humanities and Social Sciences Online, July 15, 1998. https://networks.h-net.org/h-labor.

Blackhawk, Ned. *The Rediscovery of America: Native Peoples and the Unmaking of U.S. History.* Yale University Press, 2023.

Bommersbach, Jana. "Homos on the Range." *True West*, November 1, 2005. https://truewestmagazine.com/article/homos-on-the-range.

Bourgon, Lyndsie. *Tree Thieves: Crime and Survival in North America's Woods.* Little, Brown Spark, 2022.

Brett, Mia. "If 'Secularists' Are Waging War on Christmas, Where Does That Leave the Supreme Court? It Said Christmas Trees Are Secular." *The Editorial Board*, December 17, 2021. https://www.editorialboard.com/if-secularists-are-waging-war-on-christmas-where-does-that-leave-the-supreme-court-it-said-christmas-trees-are-secular.

Briggs, Garrett W. "A Contextual Analysis of Wood-Use Behavior at Wupatki Pueblo." Master's thesis, Northern Arizona University, May 2017. ProQuest (10283007).

Brown, Alex. "For Tribes, Reforesting Means Reconnecting to History, Culture." *Stateline*, October 26, 2022. https://stateline.org/2022/10/26/for-tribes-reforesting-means-reconnecting-to-history-culture.

Bullock, Chenae. *50 Plant Medicines: Indigenous Oral History and Perspective.* Moskehtu Consulting, 2021.

Carlsen, Spike. *Woodworkers & Baseball Bats.* HarperCollins, 2009.

Cervini, Eric. *The Deviant's War: The Homosexual vs. the United States of America.* Farrar, Straus and Giroux, 2020.

Chang, Gordon H. *Ghosts of Gold Mountain: The Epic Story of the Chinese Who Built the Transcontinental Railroad.* Mariner Books, 2019.

Davis, Kimberley T., Solomon Z. Dobrowski, Philip E. Higuera, et al. "Wildfires and Climate Change Push Low-Elevation Forests Across a Critical Climate Threshold for Tree Regeneration." *Proceedings of the National Academy of Sciences* 116, no. 13 (2019): 6193–98. https://doi.org/10.1073/pnas.1815107116.

de Villiers, Marq. *The Longbow, the Schooner & the Violin: Wood and Human Achievement.* Sutherland House, 2022.

DeFonso, Chet. "The Great War and Modern Homosexuality: Transatlantic Crossings." In *Proceedings of Armistice & Aftermath: A WWI Symposium,* Michigan Tech, September 28–29, 2018. https://digitalcommons.mtu.edu/wwicc-symposium/program/program/30/.

DePillis, Lydia. "As Wildfires Grow Fiercer, Some Companies Look to Rebuild the Tree Supply Chain." *New York Times,* April 4, 2024. https://www.nytimes.com/2024/04/04/headway/wildfire-forestry-supply-chain.html.

DK. *The Tree Book: The Stories, Science, and History of Trees.* DK, 2022.

Dongoske, Kurt E., and Cindy K. Dongoske. "History in Stone: Evaluating Spanish Conversion Efforts Through Hopi Rock Art." In *Archaeologies of the Pueblo Revolt: Identity, Meaning, and Renewal in the Pueblo World,* edited by Robert W. Preucel. University of New Mexico Press, 2002.

Douglass, Frederick. *Narrative of the Life of Frederick Douglass, an American Slave.* Anti-Slavery Office, 1849. https://www.loc.gov/resource/lhbcb.25385/.

Earth Observatory. "World of Change: Burn Recovery in Yellowstone." NASA. https://earthobservatory.nasa.gov/world-of-change/Yellowstone.

Ennos, Roland. *The Age of Wood: Our Most Useful Material and the Construction of Civilization.* Scribner, 2020.

Farmer, Jared. *Trees in Paradise: A California History.* W. W. Norton, 2013.

Fedkiw, John. "The Evolving Use and Management of the Nation's Forests, Grasslands, Croplands, and Related Resources." General Technical Report RM-175. Rocky Mountain Forest and Range Experiment Station, USDA Forest Service, 1989. https://research.fs.usda.gov/treesearch/65078.

Foster, Thomas A. *Rethinking Rufus: Sexual Violations of Enslaved Men.* University of Georgia Press, 2019. https://doi.org/10.2307/j.ctv5npjgc.

Foster, Thomas A. "The Sexual Abuse of Black Men Under American Slavery." *Journal of the History of Sexuality* 20, no. 3 (2011): 445–47. https://www.jstor.org/stable/41305880.

Fry, Walter, and John Roberts White. *Big Trees*. Stanford University Press, 1930.

Fuller, Andrew S. *The Forest Tree Culturist: A Treatise on the Cultivation of American Forest Trees*. Geo. E Woodward, 1866.

Goldfarb, Ben. *Crossings: How Road Ecology Is Shaping the Future of Our Planet*. W. W. Norton, 2023.

Gooley, Tristan. *How to Read a Tree: Clues and Patterns from Bark to Leaves*. The Experiment, 2023.

Guiterman, Christopher H., Thomas W. Swetnam, and Jeffrey S. Dean. "Eleventh-Century Shift in Timber Procurement Areas for the Great Houses of Chaco Canyon." *Proceedings of the National Academy of Sciences* 113, no. 5 (2016): 1186–90. https://doi.org/10.1073/pnas.1514272112.

Hanson, Thor. *The Triumph of Seeds: How Grains, Nuts, Kernels, Pulses, and Pips Conquered the Plant Kingdom and Shaped Human History*. Basic Books, 2016.

Harlan, Chico. "A Post-Fire 'Nightmare' in New Mexico: Eight Floods in Four Weeks." *Washington Post*, July 21, 2024. https://www.washingtonpost.com /climate-environment/2024/07/21/ruidoso-new-mexico-flooding-fires-climate.

Hobbs, Kevin, and David West. *The Story of Trees: And How They Changed the World*. Laurence King, 2020.

Holten, Katie. *The Language of Trees: A Rewilding and Rewriting of Literature and Landscape*. National Geographic Books, 2023.

Hotchkiss, George W. *History of the Lumber and Forest Industry of the Northwest*. George W. Hotchkiss, 1898.

Hughes, Louis. *Thirty Years a Slave. From Bondage to Freedom. The Institution of Slavery as Seen on the Plantation and in the Home of the Planter*. South Side Printing, 1897. http://www.loc.gov/resource/lhbcb.21103.

Hunwick, Robert Foyle. "Why Christmas Is Huge in China." *Atlantic*, December 24, 2014. https://www.theatlantic.com/international/archive/2014/12/why -christmas-is-huge-in-china/384040.

Hurston, Zora Neale. *Barracoon: The Story of the Last "Black Cargo."* HarperCollins, 2018.

Hyman, Harold M. *Soldiers and Spruce: Origins of the Loyal Legion of Loggers and Lumbermen*. Institute of Industrial Relations, 1963.

Ingraham, J. H. *The Sunny South; or, the Southerner at Home, Embracing Five Years' Experience of a Northern Governess in the Land of the Sugar and the Cotton*. G. G. Evans, 1860. https://www.loc.gov/item/rc01002308.

Johnson, Colin R. "Camp Life: The Queer History of 'Manhood' in the Civilian Conservation Corps, 1933–1937." *American Studies* 48, no. 2, (2007): 19–35. https://www.jstor.org/stable/40644066.

Keegan, Charles E., Collin B. Sorenson, Todd A. Morgan, Jean M. Daniels, and
Steven W. Hayes. "Impact of the Great Recession on the Forest Products
Industry in the Western United States." In *Moving from Status to Trends:
Forest Inventory and Analysis Symposium 2012, GTR-NRS-P-105*, US For-
est Service, Pacific Northwest Research Station. 1–8. https://research.fs.usda
.gov/treesearch/42663.

Keltner, Robert W. "Tar Paper Shacks in Arcadia: Housing for Ethnic Minority
Groups in the Company Town of Bauxite, Arkansas." *Arkansas Historical
Quarterly* 60, no. 4 (2001): 341–59. https://doi.org/10.2307/40038252.

Kinsey, Alfred C., Wardell B. Pomeroy, and Clyde E. Martin. *Sexual Behavior in
the Human Male*. W. B. Saunders, 1948.

Koberstein, Paul, and Jessica Applegate. *Canopy of Titans: The Life and Times of
the Great North American Temperate Rainforest*. OR Books, 2022.

Lee, Chloe. "Ancient Wisdom for Modern Challenges: The Influence of Greek
Philosophers on Environmental Sustainability." *Journal of Student Research*
12, no. 4 (2023). https://doi.org/10.47611/jsr.v12i4.2217.

Lee, KangJae Jerry, Mariela Fernandez, David Scott, and Myron Floyd. "Slow
Violence in Public Parks in the U.S.: Can We Escape Our Troubling Past?"
Social & Cultural Geography 24, no. 7 (2022): 1185–1202. https://doi.org/10
.1080/14649365.2022.2028182.

Leopold, Aldo. "The Last Stand of the Wilderness." *American Forests and
Forest Life 31, no. 382 (1925)*.

Lewis, Amanda. *Tracking Giants: Big Trees, Tiny Triumphs, and Misadventures
in the Forest*. Greystone Books, 2023.

Library of Congress. "The First Peoples of California." California as I Saw It:
First-Person Narratives of California's Early Years, 1849 to 1900. https://
www.loc.gov/collections/california-first-person-narratives/articles-and
-essays/early-california-history/first-peoples-of-california.

Liu, Weijia, Zhi Qu, Xuemei Wang, et al. "Microplastic Exposure Is Associated
with Male Reproductive Health." *Medical Review* 4, no. 6 (2024): 549–52.
https://doi.org/10.1515/mr-2024-0069.

Logan, William Bryant. *Sprout Lands: Tending the Endless Gift of Trees*. W. W.
Norton, 2019.

Lowe, Gary D. *Debunking the Sequoia Honoring Sequoyah Myth*. Stanford
Digital Library, 2018.

Lowe, Gary D. "Endlicher's Sequence: The Naming of the Genus Sequoia." *Fre-
montia* 40, no. 1 (2012): 25.

Loyal Legion of Loggers and Lumbermen. "How Spruce Became a Forest King."
Monthly Bulletin, March 1918.

MacCleery, Douglas W. *American Forests: A History of Resiliency and Recovery.* Forest History Society, 2011. https://foresthistory.org/wp-content/uploads /2016/12/American_Forests.pdf.

MacKinnon, J. B. *The Once and Future World: Nature as It Was, as It Is, as It Could Be.* Houghton Mifflin Harcourt, 2013.

Mancuso, Stefano. *Planting Our World.* Other Press, 2023.

Martin, Claire. "Scientists See Insect Outbreaks From Space." *Smithsonian*, February 1, 2013. https://www.smithsonianmag.com/science-nature/scientists -see-insect-outbreaks-from-space-10365497.

Mason, Daniel. *North Woods: A Novel.* Random House, 2023.

Mason, Laura. *Pine.* Reaktion Books, 2013.

May, Robert E. *Yuletide in Dixie: Slavery, Christmas, and Southern Memory.* University of Virginia Press, 2019.

Melendez, Bill, dir. *A Charlie Brown Christmas.* Written by Charles M. Schulz. First aired December 9, 1965, on CBS.

Muir, John. *My First Summer in the Sierra.* Houghton Mifflin, 1911.

Muir, John. *Steep Trails.* Houghton Mifflin, 1918.

Mytting, Lars. *Norwegian Wood: Chopping, Stacking, and Drying Wood the Scandinavian Way.* Abrams, 2015.

Nisbet, Jack. *The Collector: David Douglas and the Natural History of the Northwest.* Sasquatch Books, 2009.

Oregon Forest Resources Institute. *Oregon Forest Facts 2023–24 Edition.* https:// oregonforests.org/media/995.

Parks, Sean A., Christopher H. Guiterman, Ellis Q. Margolis, et al. "A Fire Deficit Persists Across Diverse North American Forests Despite Recent Increases in Area Burned." *Nature Communications* 16 (2025): 1493. https://doi .org/10.1038/s41467-025-56333-8.

Pastor, John. *White Pine: The Natural and Human History of a Foundational American Tree.* Island Press, 2023.

Perlin, John. *The Forest Journey: The Role of Trees in the Fate of Civilization.* Patagonia, 2023.

Perry, Imani. *South to America: A Journey Below the Mason-Dixon to Understand the Soul of a Nation.* Ecco, 2021.

Pollan, Michael. *The Omnivore's Dilemma: A Natural History of Four Meals.* Penguin, 2007.

Pollan, Michael. "The Triumph of Burbopolis." *New York Times Magazine*, April 9, 2000. https://www.nytimes.com/2000/04/09/magazine/the-triumph -of-burbopolis.html.

Powers, Richard. *The Overstory: A Novel.* W. W. Norton, 2018.

Preston, Richard. *The Wild Trees: A Story of Passion and Daring.* Random House, 2007.

Proulx, Annie. *Barkskins: A Novel.* Simon & Schuster, 2016.

Proulx, Annie. *Fen, Bog and Swamp: A Short History of Peatland Destruction and Its Role in the Climate Crisis.* Simon & Schuster, 2022.

Prouty, Andrew Mason. *More Deadly Than War! Pacific Coast Logging, 1827–1981.* Garland, 1985.

Purdy, Jedediah. *After Nature: A Politics for the Anthropocene.* Harvard University Press, 2015.

Purdy, Jedediah. "Environmentalism's Racist History." *New Yorker,* August 13, 2015. https://www.newyorker.com/news/news-desk/environmentalisms-racist-history.

Rawlence, Ben. *The Treeline: The Last Forest and the Future of Life on Earth.* St. Martin's Griffin, 2023.

Raworth, Kate. *Doughnut Economics: Seven Ways to Think Like a 21st-Century Economist.* Chelsea Green, 2018.

Ritchie, Hannah. *Not the End of the World: How We Can Be the First Generation to Build a Sustainable Planet.* Little, Brown Spark, 2024.

Roa, Michael. "Humans in California's Coniferous Forests." In *The Conifer Connection.* California Community Forests Foundation and California State Parks, 2011. https://www.parks.ca.gov/?page_id=26781.

Robbins, Jim. "To Protect Giant Sequoias, They Lit a Fire." *New York Times,* July 9, 2024. https://www.nytimes.com/2024/07/09/science/redwoods-wildfires-indigenous-tribes-california.html.

Roos, Christopher I., Christopher H. Guiterman, Ellis Q. Margolis, et al. "Indigenous Fire Management and Cross-Scale Fire-Climate Relationships in the Southwest United States from 1500 to 1900 CE." *Science Advances* 8, no. 49 (2022): eabq3221. https://doi.org/10.1126/sciadv.abq3221.

Rutkow, Eric. *American Canopy: Trees, Forests, and the Making of a Nation.* Simon & Schuster, 2012.

Schlanger, Zoë. *The Light Eaters: How the Unseen World of Plant Intelligence Offers a New Understanding of Life on Earth.* Harper, 2024.

Schmit, Dean. "'They Were Men in Those Days': Gender, Class, and Ethnicity in the Paul Bunyan Tales." Honors capstone project, University of Minnesota, 2023. https://digitalcommons.morris.umn.edu/honors.

Seneca. *The Tragedies of Seneca.* Translated by Frank Justus Miller. University of Chicago Press, 1907. https://www.gutenberg.org/files/57999/57999-h/57999-h.htm.

Shapiro, Laurie Gwen. "He Bombed the Nazis, Outwitted the Soviets and Modernized Christmas." *New York Times,* December 17, 2021. https://www.nytimes.com/2021/12/17/nyregion/bomber-pilot-christmas-trees.html.

Shashkevich, Alex. "New Collection at Stanford Libraries Offers Extensive Materials on Discovery, Exhibitions of Giant Sequoia Trees." *Stanford Report*, March 9, 2017. https://news.stanford.edu/2017/03/09/stanford-libraries-obtain-collection-giant-sequoias.

Silkenat, David. *Scars on the Land: An Environmental History of Slavery in the American South*. Oxford University Press, 2022.

Słupianek, Aleksandra, Alicja Dolzblasz, and Katarzyna Sokołowska. "Xylem Parenchyma—Role and Relevance in Wood Functioning in Trees." *Plants* 10, no. 6 (2021): 1247. https://doi.org/10.3390/plants10061247.

Snyder, Christina. "Indian Slavery." In *Oxford Research Encyclopedia of American History*. Oxford University Press, 2014. https://doi.org/10.1093/acrefore/9780199329175.013.5.

Spencer, Betty Goodwin. *The Big Blowup*. Caxton Printers, 1956.

St. George, Zach. "They Seem to Be Immortal." *Guernica*, April 15, 2015. https://www.guernicamag.com/they-seem-to-be-immortal/.

St. George, Zach. *The Journeys of Trees: A Story About Forests, People, and the Future*. W. W. Norton, 2020.

Stewart, Amy. *Flower Confidential: The Good, the Bad, and the Beautiful*. Algonquin Books, 2008.

Strickland, William. *Journal of a Tour in the United States of America, 1794–95*, edited by J. E. Strickland. New-York Historical Society, 1971.

Taiz, Lincoln, and Lee Taiz. *Flora Unveiled: The Discovery and Denial of Sex in Plants*. Oxford University Press, 2017.

Tonsfeldt, Ward. *The U.S. Army Spruce Production Division at Vancouver Barracks, Washington, 1917–1919*. East Slope Cultural Services, 2013. https://www.npshistory.com/publications/fova/army-spruce-production.pdf.

Trouet, Valerie. *Tree Story: The History of the World Written in Rings*. Johns Hopkins University Press, 2020.

Tudge, Colin. *The Tree: A Natural History of What Trees Are, How They Live, and Why They Matter*. Crown, 2006.

Turner, Nancy Chapman, and Marcus A. M. Bell. "The Ethnobotany of the Coast Salish Indians of Vancouver Island." *Economic Botany* 25, no. 1 (1971): 63–104. https://www.jstor.org/stable/4253212.

Twain, Mark. *Life on the Mississippi*. Harper, 1899. https://www.loc.gov/item/99005381/.

Tweed, William C. *King Sequoia: The Tree That Inspired a Nation, Created Our National Park System, and Changed the Way We Think About Nature*. Heyday, 2015.

Vaillant, John. *Fire Weather: On the Front Lines of a Burning World*. Vintage, 2024.

Vaillant, John. *The Golden Spruce: A True Story of Myth, Madness, and Greed.* W. W. Norton, 2006.

Vaughan, Carson. "A 'Green Glacier' Is Dismantling the Great Plains." *New York Times*, December 7, 2023. https://www.nytimes.com/2023/12/07/opinion /prairie-great-plains-trees.html.

Vietze, Andrew. *White Pine: American History and the Tree That Made a Nation.* Globe Pequot Press, 2017.

Watson, Adam S. "Long-Distance Wood Procurement and the Chaco Florescence." *Proceedings of the National Academy of Sciences* 113, no. 5 (2016): 1118–20. https://doi.org/10.1073/pnas.1521904113.

Waxman, Olivia B. "How Christmas Trees Became a Holiday Tradition." *Time*, December 5, 2019. https://time.com/5736523/history-of-christmas-trees.

Wells, V. S. "The Secret Queer History of BC's Wild West." *Georgia Straight*, January 18, 2024. https://www.straight.com/city-culture/secret-queer-history -of-bcs-wild-west.

Weyerhaeuser. "Questions and Answers About Wood Adhesives and Formaldehyde Emissions." Technical Resource Sheet, December 2023. https://www .weyerhaeuser.com/application/files/3017/0300/4831/TRS-1505.pdf.

White, Lynn, Jr. "The Historical Roots of Our Ecologic Crisis." *Science* 155, no. 3767 (1967): 1203–07. https://www.science.org/doi/10.1126/science .155.3767.1203.

Williams, Gerald W. "The Spruce Production Division." *Forest History Today*, Spring/Fall 2018. https://foresthistory.org/wp-content/uploads /2019/06/9-Williams_Spruce_Division.pdf.

Wilson, James A., Robert S. Weiner, Joseph P. Carzoli, and Rodger Kram. "Were Timbers Transported to Chaco Using Tumplines? A Feasibility Study." *Journal of Archaeological Science: Reports* 48 (2023): 103876. https://doi .org/10.1016/j.jasrep.2023.103876.

Wolfe, Daniel, and Aaron Steckelberg. "A Surprising Byproduct of Wildfires: Contaminated Drinking Water." *Washington Post*, July 29, 2024. https:// www.washingtonpost.com/climate-environment/interactive/2024/wildfire -drinking-water-supply-danger.

Woodford, James. *The Wollemi Pine: The Incredible Discovery of a Living Fossil from the Age of the Dinosaurs.* Text Publishing, 2012.

Zeidler, Aleš, Vlastimil Borůvka, and Ondřej Schönfelder. "Comparison of Wood Quality of Douglas Fir and Spruce from Afforested Agricultural Land and Permanent Forest Land in the Czech Republic." *Forests* 9, no. 1 (2018): 13. https://doi.org/10.3390/f9010013.

RAISING READERS
Books Build Bright Futures

Thank you for reading this book and for being a reader of books in general. We are so grateful to share being part of a community of readers with you, and we hope you will join us in passing our love of books on to the next generation of readers.

Did you know that reading for enjoyment is the single biggest predictor of a child's future happiness and success?

More than family circumstances, parents' educational background, or income, reading impacts a child's future academic performance, emotional well-being, communication skills, economic security, ambition, and happiness.

Studies show that kids reading for enjoyment in the US is in rapid decline:

- In 2012, 53% of 9-year-olds read almost every day. Just 10 years later, in 2022, the number had fallen to 39%.
- In 2012, 27% of 13-year-olds read for fun daily. By 2023, that number was just 14%.

Together, we can commit to **Raising Readers** and change this trend. How?

- Read to children in your life daily.
- Model reading as a fun activity.
- Reduce screen time.
- Start a family, school, or community book club.
- Visit bookstores and libraries regularly.
- Listen to audiobooks.
- Read the book before you see the movie.
- Encourage your child to read aloud to a pet or stuffed animal.
- Give books as gifts.
- Donate books to families and communities in need.

BOB1217

Books build bright futures, and **Raising Readers** is our shared responsibility.

For more information, visit **JoinRaisingReaders.com**

Sources: National Endowment for the Arts, National Assessment of Educational Progress, WorldBookDay.com, Nielsen BookData's 2023 "Understanding the Children's Book Consumer"